MARTINI'S

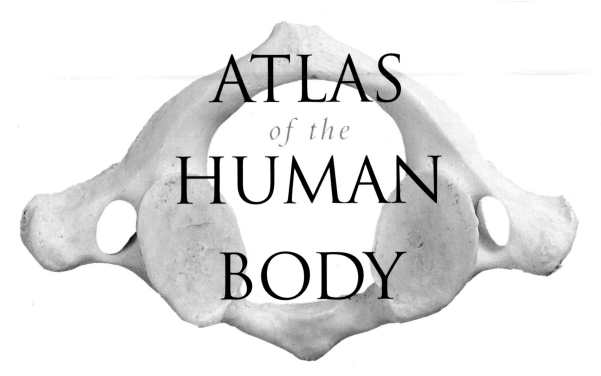

ATLAS

of the

HUMAN

BODY

by

FREDERIC H. MARTINI, PH.D.

University of Hawaii

with

WILLIAM C. OBER, M.D.
Art Coordinator and Illustrator

KATHLEEN WELCH
Clinical Consultant

CLAIRE W. GARRISON, R.N.
Illustrator

RALPH T. HUTCHINGS, PH.D.
Biomedical Photographer

PEARSON

Benjamin
Cummings

San Francisco Boston New York
Capetown Hong Kong London Madrid Mexico City
Montreal Munich Paris Singapore Sidney Tokyo Toronto

Acquisitions Editor: Leslie Berriman
Project Editor: Katy German
Art Development Editor: Blake Kim
Editorial Assistant: Kelly Reed
Managing Editor: Deborah Cogan
Production Supervisor: Caroline Ayres
Production Management: Carlisle Publishing Services
Compositor: Carlisle Publishing Services
Art Coordinator: Jean Lake

Interior Designer: Emily Friel
Cover Designer: Yvo Riezebos
Illustrators: Precision Graphics
Director, Image Resource Center: Melinda Patelli
Image Rights and Permissions Manager: Zina Arabia
Manufacturing Buyer: Stacey Weinberger
Executive Marketing Manager: Gordon Lee
Text printer: R.R. Donnelley, Willard
Cover printer: Phoenix Color

Cover Photo Credit: Ralph Hutchings

Photo Credits: 1a, b, 2b, 3c, d, 4a-c, 5a-f, 7a-d, 8a-d, 9a-c, 10, 11a, c, 12a, 14a-d, 15a, c, 16a-c, 18a-c, 19, 20a, 21a-e, 22a-c, 23a, b, 24a, d, 25, 26a, b, 27d, 28, 29a-c, 31, 32, 33b, d, 34a-d, 35a, b, e, f, 36a, b, 37b, 38a, d-f, 39c, 40b, 41a, 42a, b, 43a-d, 44a, b, 45b, d, 46a, b, 49b-e, 51a-d, 52, 53a, 54a-c, 55a, b, 57a, b, 60a, b, 61a-61c, 63, 64, 65, 66, 68b, c, 70a, b, 72a, 73a, b, 74, 75a-d, 76b, 77, 79a, b, 80a, b, 81b, 82b, 83a, b, 84a, b, 85b, 86a, b, 87b, 88b, 89, 90a, b Ralph T. Hutchings 2a, 13f, 23c, 35c, 35g, 38b, 78a Patrick M. Timmons/Michael J. Timmons 3a, 17, 27a-c, 30, 33c, 36, 37a, 39a, d, 40a, 68a, 69a, 76a, 81a, 82a, 87a Mentor Networks, Inc. 6 Image provided by The Digital Cadaver™ Project, courtesy of Visible Productions, Inc. 11b, 12b-d, 13b-e, 20b, 56a-c, 58a-c, 72b, 78b-g, 86c, 87c Frederic H. Martini, Inc. 13a, Pat Lynch/Photo Researchers, Inc. 13g, 24c, 33a, 39b, 54d, 59, 69b, 85a Custom Medical Stock Photo, Inc. 15b Martin M. Rotker 35d Thiem Verlagsgruppe 38c, 47a-d, 49a, 62a, b Pearson Education, PH College 42c, 44c Wellcome Trust Medical 42d, 67 CNRI/Science Photo Library/Photo Researchers, Inc. Researchers, Inc. 45a David York/Medichrome/The Stock Shop, Inc. 48b Marconi Medical Systems, Inc. 48c, 50a, 71b, 78i www.NetAnatomy.com 50b ISM/Phototake NYC 50c, 53d Christopher J. Bodin, M. D., Tulane University Medical Center 53c Barry Slaven/P. Mode Photography/Photo Researchers, Inc. 53e P.M. Motta, A. Caggiati, G. Macchiarelli/Science Photo Library/Photo Researchers, Inc. 45c Science Photo Library/Photo Researchers, Inc. Researchers, Inc. 56d-f, 58d-f Visible Human Project/National Institutes of Health, National Library of Medicine 71a L. Basset/Visuals Unlimited 78h Michael L. Richardson; University of Washington of School of Medicine, Dept. of Radiology; www.rad.washington.edu

Library of Congress Cataloging-in-Publication Data

Martini, Frederic.
   Martini's atlas of the human body/by Frederic H. Martini; with William C. Ober, art coordinator and illustrator... [et al.].—
8th ed.
      p. ; cm.
ISBN-13: 978-0-321-50597-2
ISBN-10: 0-321-50597-2
   1. Human anatomy—Atlases. I. Ober, William C. II. Title III. Title: Atlas of the human body
   [DNLM: 1. Anatomy—Atlases. 2. Physiology—Atlases. QS 17 M386m 2009]
   QM25.M286 2009
   611—dc22   2007039454

1 2 3 4 5 6 7 8 9 10-DOW-12 11 10 09 08

PEARSON
Benjamin
Cummings

www.aw-bc.com

ISBN 10: 0-321-50597-2
ISBN 13: 978-0-321-50597-2

For students in an introductory level anatomy and physiology course, an interpretive anatomical illustration is often the best way to introduce important information without distracting clutter. However, for true understanding, it is important to relate that interpretive view to "real world" anatomy, which is usually much more complex. For example, it is easier to learn the distribution of the major arteries in images that show only arterial and skeletal structures, and yet the concepts learned in that way can be hard to apply and interpret when dealing with a cadaver dissection or a hospital emergency. You might initially think that the best solution would be for a textbook to present illustrations in pairs—an interpretive drawing in conjunction with a matching "real" view such as a cadaver photo or medical scan. In fact that is how many figures in my textbooks are organized. But to do that globally, for all structures and all systems, would be impractical and unwieldy. First of all, the same photographs would have to appear multiple times—the same image of the dissection of the forearm would be paired with illustrations dealing with the bones, muscles, nerves, arteries, veins, and lymphatics of the forearm. Even if this redundancy were tolerable, the textbook would double in length unless the illustrations were reduced in size by half—an unacceptable option because larger images enhance understanding.

This *Atlas* solves that dilemma and makes it easy to relate clear diagrammatic illustrations in the textbook to the real world of dissection images and medical scans. It has been designed to supplement the anatomical illustrations in *Fundamentals of Anatomy & Physiology,* Eighth Edition. Figure captions in that textbook indicate the related numbered plates in this *Atlas.* Together, that textbook and this *Atlas* provide a comprehensive and highly-visual orientation to the anatomy of the human body.

Because it contains such a comprehensive series of images, this *Atlas* is also a useful companion for my other textbooks, including: Martini/Timmons/Tallitsch *Human Anatomy;* Martini/ Bartholomew *Essentials of Anatomy & Physiology;* and Martini *Anatomy & Physiology*—all available from Benjamin Cummings.

The images in this *Atlas* are regionally organized. In key areas of the body, diagnostic medical images are paired with cadaver views. This will help you take an additional step—first from the diagrammatic illustration (in the textbook) to a cadaver dissection in a laboratory setting, and then from that inert cadaver to a medical image of a living person.

This *Atlas* also contains a visual summary of human embryological development. The timing and depth of coverage of this topic vary widely across classrooms; some instructors cover the embryological development of individual body systems as they discuss each system, whereas others use the topic to wrap up the end of the course, or address the topic briefly only at the beginning and end of the course. There is general agreement, however, that understanding the key events in anatomical development requires an ability to visualize events happening in three-dimensional space. So in the Embryology Summary section of the *Atlas,* we have created a visual presentation that pairs each written description with interpretive art. You can refer to the sub-sections in the Embryology Summary, identified by topic and body system, as needed as your course proceeds.

In the preparation of this *Atlas* I have been very fortunate to work with Ralph Hutchings and Ms. Claire Garrison, considered by many to be the world's preeminent anatomical photographer, and Dr. William Ober, a superb and creative medical illustrators who worked with me on the complexities of labeling the plates and helped create the Embryology Summaries. I would also like to thank the faculty who contacted me over the last three years with suggestions for the reorganization and improvement of this *Atlas.*

I hope you find this *Atlas* useful in your studies. If you have comments or suggestions for improvement, please contact me at the address below.

Frederic H. Martini
martini@maui.net

Contents

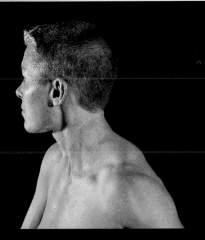 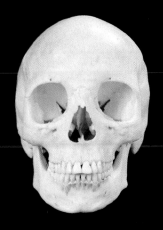 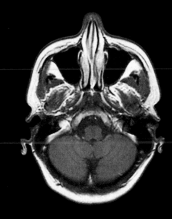

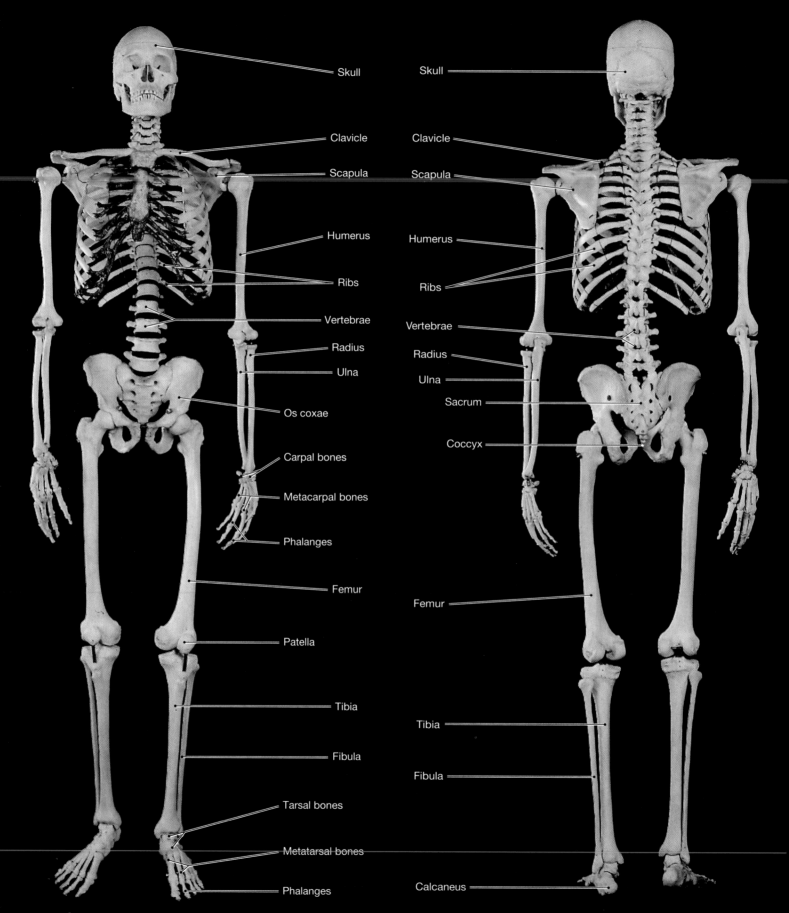

Skull

Clavicle

Scapula

Humerus

Ribs

Vertebrae

Radius

Ulna

Os coxae

Carpal bones

Metacarpal bones

Phalanges

Femur

Patella

Tibia

Fibula

Tarsal bones

Metatarsal bones

Phalanges

Skull

Clavicle

Scapula

Humerus

Ribs

Vertebrae

Radius

Ulna

Sacrum

Coccyx

Femur

Tibia

Fibula

Calcaneus

PLATE **1a** THE SKELETON, ANTERIOR VIEW

PLATE **1b** THE SKELETON, POSTERIOR VIEW

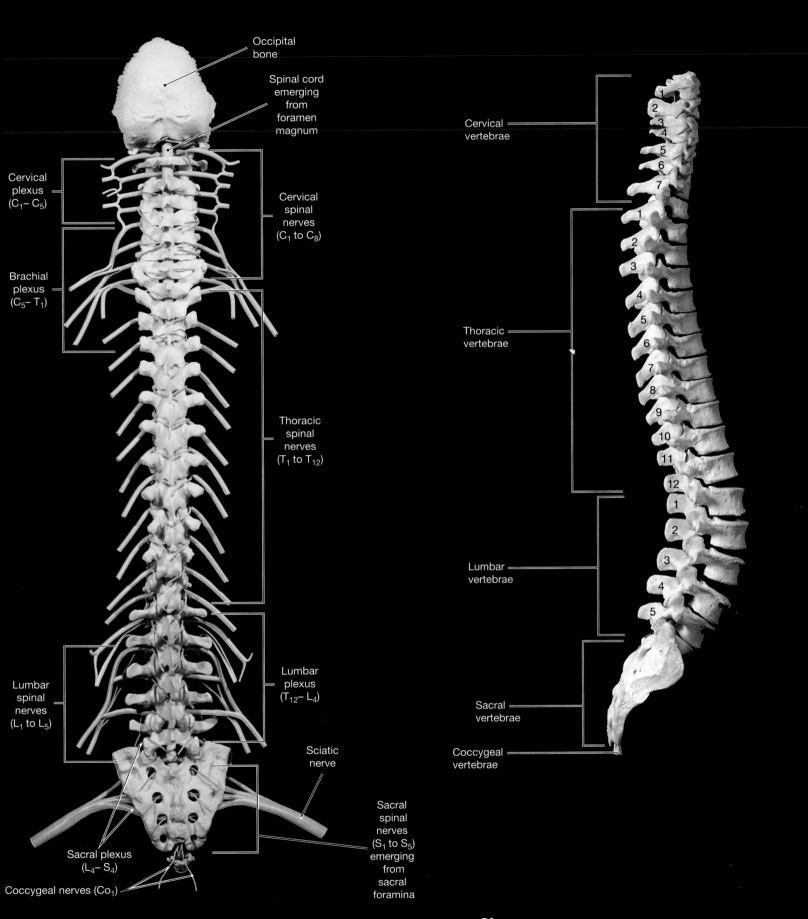

Occipital bone

Spinal cord emerging from foramen magnum

Cervical plexus (C_1– C_5)

Brachial plexus (C_5– T_1)

Lumbar spinal nerves (L_1 to L_5)

Sacral plexus (L_4– S_4)

Coccygeal nerves (Co_1)

Cervical spinal nerves (C_1 to C_8)

Thoracic spinal nerves (T_1 to T_{12})

Lumbar plexus (T_{12}– L_4)

Sciatic nerve

Sacral spinal nerves (S_1 to S_5) emerging from sacral foramina

Cervical vertebrae

Thoracic vertebrae

Lumbar vertebrae

Sacral vertebrae

Coccygeal vertebrae

PLATE **2a** THE VERTEBRAL COLUMN AND SPINAL NERVES

PLATE **2b** THE VERTEBRAL COLUMN, LATERAL VIEW

PLATE **3a**   SURFACE ANATOMY OF THE
HEAD AND NECK, ANTERIOR VIEW

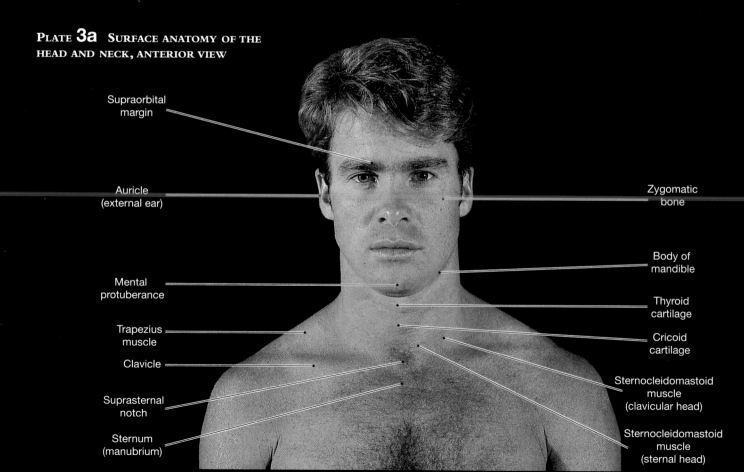

Supraorbital
margin

Auricle
(external ear)

Zygomatic
bone

Body of
mandible

Mental
protuberance

Thyroid
cartilage

Trapezius
muscle

Cricoid
cartilage

Clavicle

Sternocleidomastoid
muscle
(clavicular head)

Suprasternal
notch

Sternum
(manubrium)

Sternocleidomastoid
muscle
(sternal head)

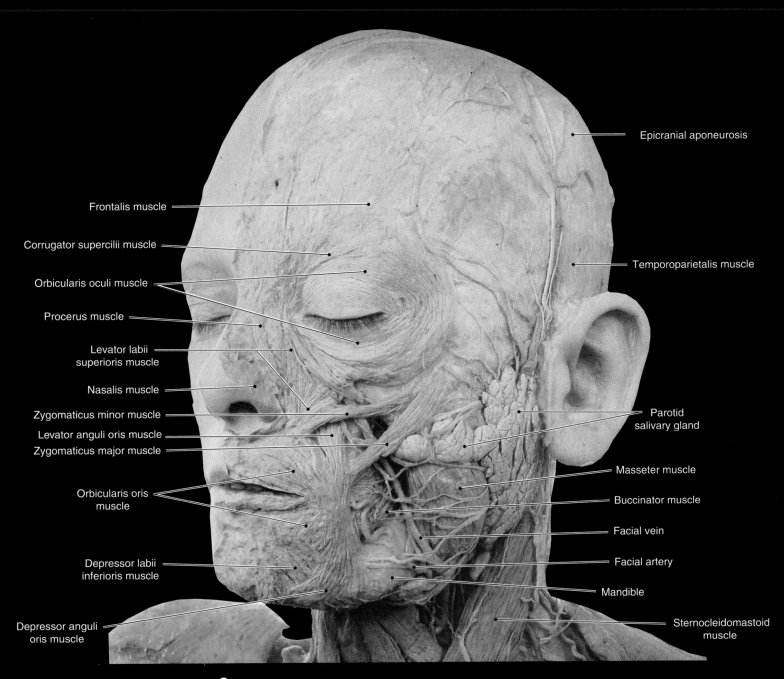

Epicranial aponeurosis

Frontalis muscle

Corrugator supercilii muscle

Temporoparietalis muscle

Orbicularis oculi muscle

Procerus muscle

Levator labii
superioris muscle

Nasalis muscle

Zygomaticus minor muscle

Parotid
salivary gland

Levator anguli oris muscle

Zygomaticus major muscle

Masseter muscle

Orbicularis oris
muscle

Buccinator muscle

Facial vein

Depressor labii
inferioris muscle

Facial artery

Mandible

Depressor anguli
oris muscle

Sternocleidomastoid
muscle

PLATE **3c** Superficial dissection of the face, anterolateral view

PLATE 3d SUPERFICIAL
DISSECTION OF THE FACE,
LATERAL VIEW

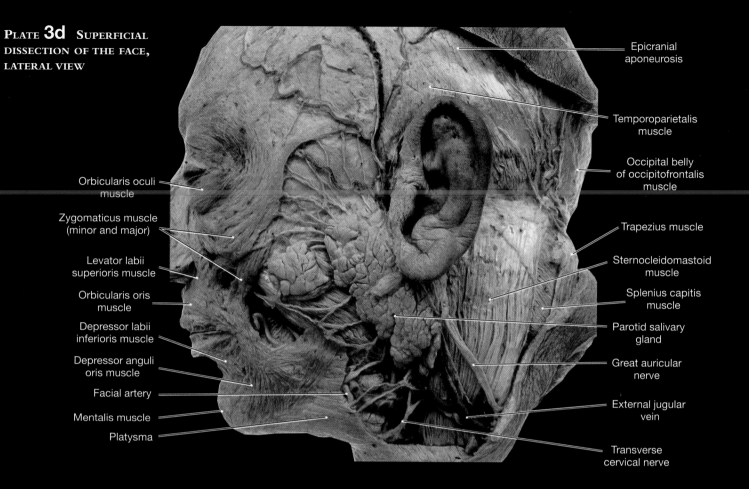

Orbicularis oculi
muscle

Zygomaticus muscle
(minor and major)

Levator labii
superioris muscle

Orbicularis oris
muscle

Depressor labii
inferioris muscle

Depressor anguli
oris muscle

Facial artery

Mentalis muscle

Platysma

Epicranial
aponeurosis

Temporoparietalis
muscle

Occipital belly
of occipitofrontalis
muscle

Trapezius muscle

Sternocleidomastoid
muscle

Splenius capitis
muscle

Parotid salivary
gland

Great auricular
nerve

External jugular
vein

Transverse
cervical nerve

PLATE 4a PAINTED SKULL, LATERAL VIEW

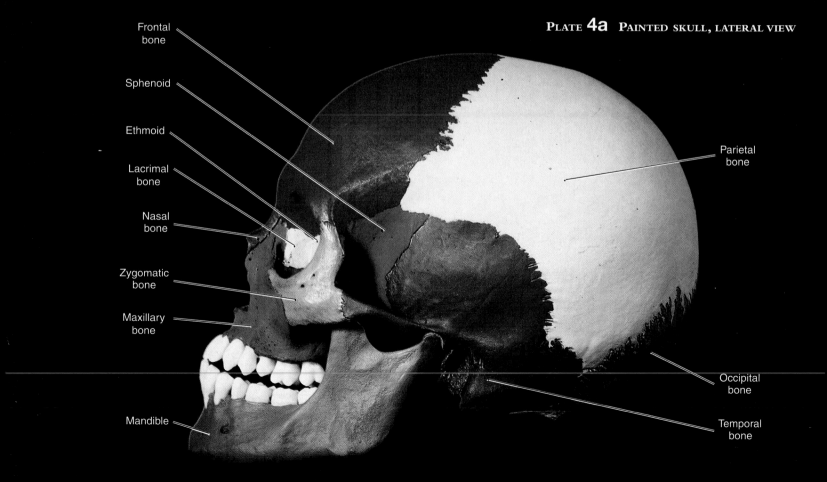

Frontal
bone

Sphenoid

Ethmoid

Lacrimal
bone

Nasal
bone

Zygomatic
bone

Maxillary
bone

Mandible

Parietal
bone

Occipital
bone

Temporal
bone

PLATE **4b** PAINTED SKULL,
ANTEROLATERAL VIEW

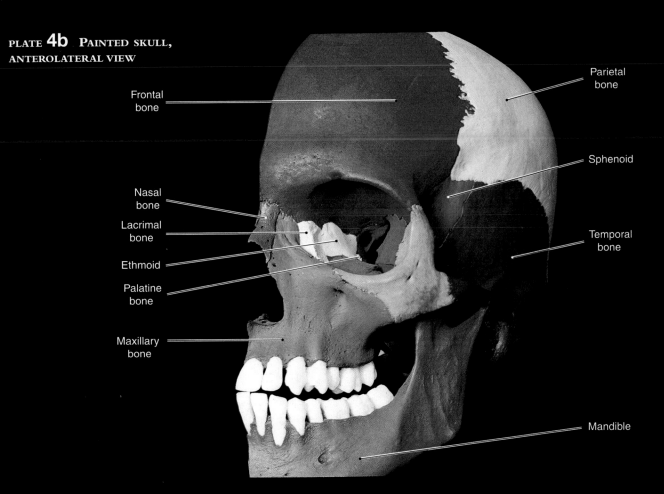

Frontal
bone

Parietal
bone

Sphenoid

Nasal
bone

Lacrimal
bone

Ethmoid

Palatine
bone

Temporal
bone

Maxillary
bone

Mandible

PLATE **4c** PAINTED SKULL,
MEDIAL VIEW

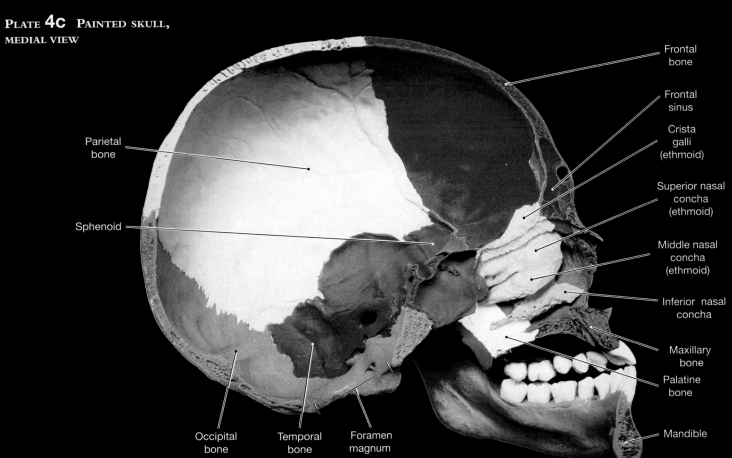

Frontal
bone

Frontal
sinus

Crista
galli
(ethmoid)

Parietal
bone

Superior nasal
concha
(ethmoid)

Sphenoid

Middle nasal
concha
(ethmoid)

Inferior nasal
concha

Maxillary
bone

Palatine
bone

Mandible

Occipital
bone

Temporal
bone

Foramen
magnum

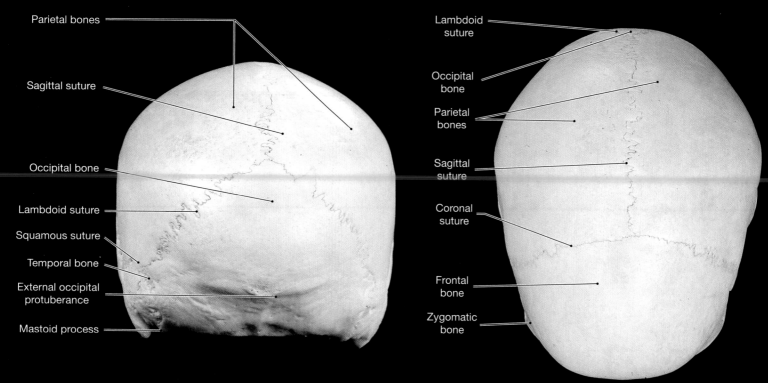

Parietal bones

Sagittal suture

Occipital bone

Lambdoid suture

Squamous suture

Temporal bone

External occipital protuberance

Mastoid process

Lambdoid suture

Occipital bone

Parietal bones

Sagittal suture

Coronal suture

Frontal bone

Zygomatic bone

PLATE 5a ADULT SKULL, POSTERIOR VIEW

PLATE 5b ADULT SKULL, SUPERIOR VIEW

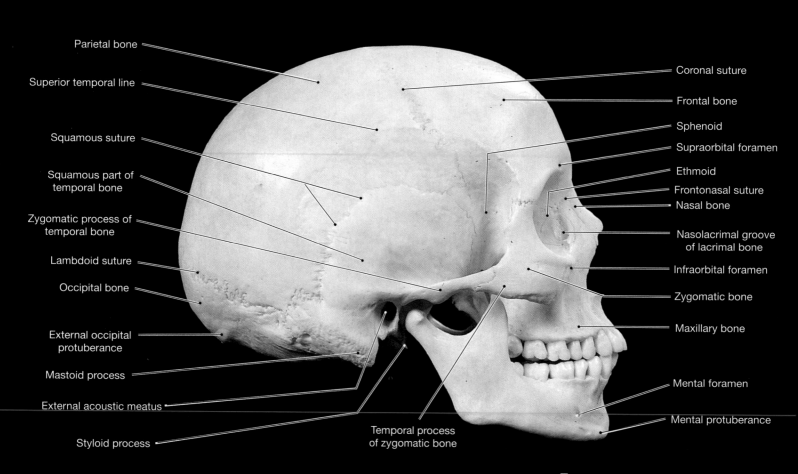

Parietal bone

Superior temporal line

Squamous suture

Squamous part of temporal bone

Zygomatic process of temporal bone

Lambdoid suture

Occipital bone

External occipital protuberance

Mastoid process

External acoustic meatus

Styloid process

Coronal suture

Frontal bone

Sphenoid

Supraorbital foramen

Ethmoid

Frontonasal suture

Nasal bone

Nasolacrimal groove of lacrimal bone

Infraorbital foramen

Zygomatic bone

Maxillary bone

Mental foramen

Mental protuberance

Temporal process of zygomatic bone

PLATE 5c ADULT SKULL, LATERAL VIEW

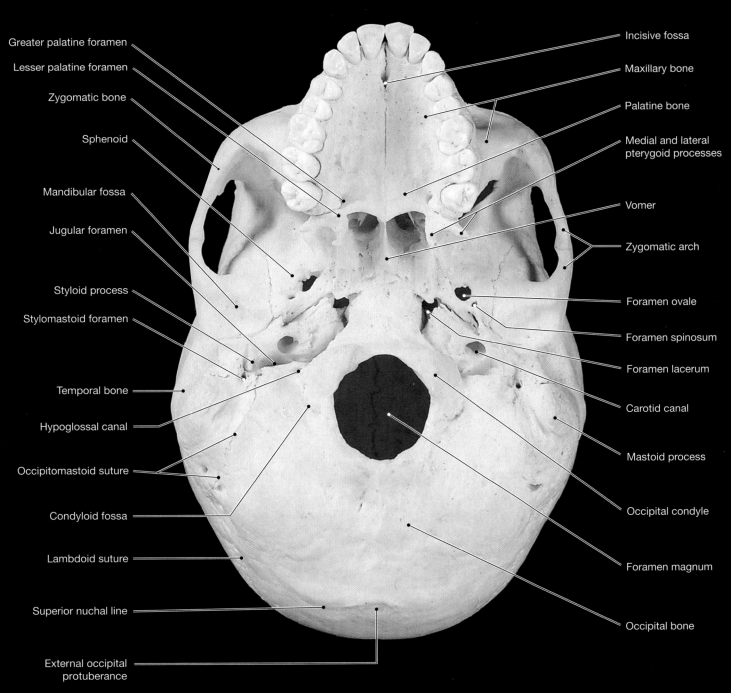

Greater palatine foramen

Lesser palatine foramen

Zygomatic bone

Sphenoid

Mandibular fossa

Jugular foramen

Styloid process

Stylomastoid foramen

Temporal bone

Hypoglossal canal

Occipitomastoid suture

Condyloid fossa

Lambdoid suture

Superior nuchal line

External occipital
protuberance

Incisive fossa

Maxillary bone

Palatine bone

Medial and lateral
pterygoid processes

Vomer

Zygomatic arch

Foramen ovale

Foramen spinosum

Foramen lacerum

Carotid canal

Mastoid process

Occipital condyle

Foramen magnum

Occipital bone

PLATE **5d**   ADULT SKULL, INFERIOR VIEW (MANDIBLE REMOVED)

PLATE 5e ADULT SKULL, ANTERIOR VIEW

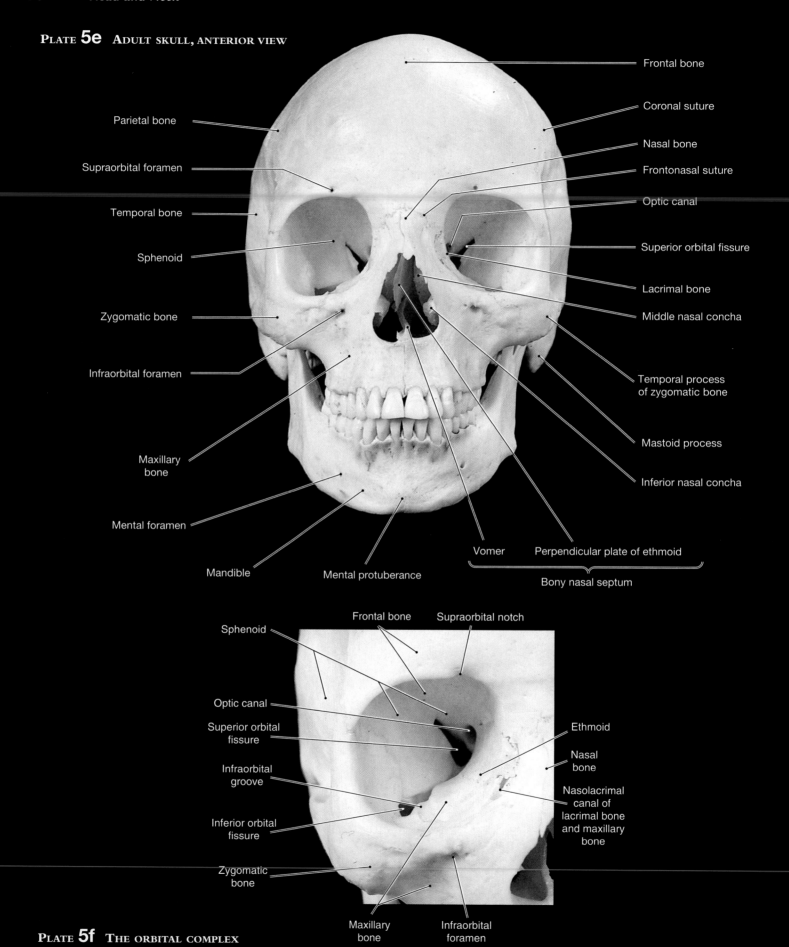

Frontal bone

Coronal suture

Nasal bone

Frontonasal suture

Optic canal

Superior orbital fissure

Lacrimal bone

Middle nasal concha

Temporal process
of zygomatic bone

Mastoid process

Inferior nasal concha

Parietal bone

Supraorbital foramen

Temporal bone

Sphenoid

Zygomatic bone

Infraorbital foramen

Maxillary
bone

Mental foramen

Mandible

Mental protuberance

Vomer   Perpendicular plate of ethmoid

Bony nasal septum

Frontal bone   Supraorbital notch

Sphenoid

Optic canal

Superior orbital
fissure

Infraorbital
groove

Inferior orbital
fissure

Zygomatic
bone

Ethmoid

Nasal
bone

Nasolacrimal
canal of
lacrimal bone
and maxillary
bone

Maxillary
bone

Infraorbital
foramen

PLATE 5f THE ORBITAL COMPLEX

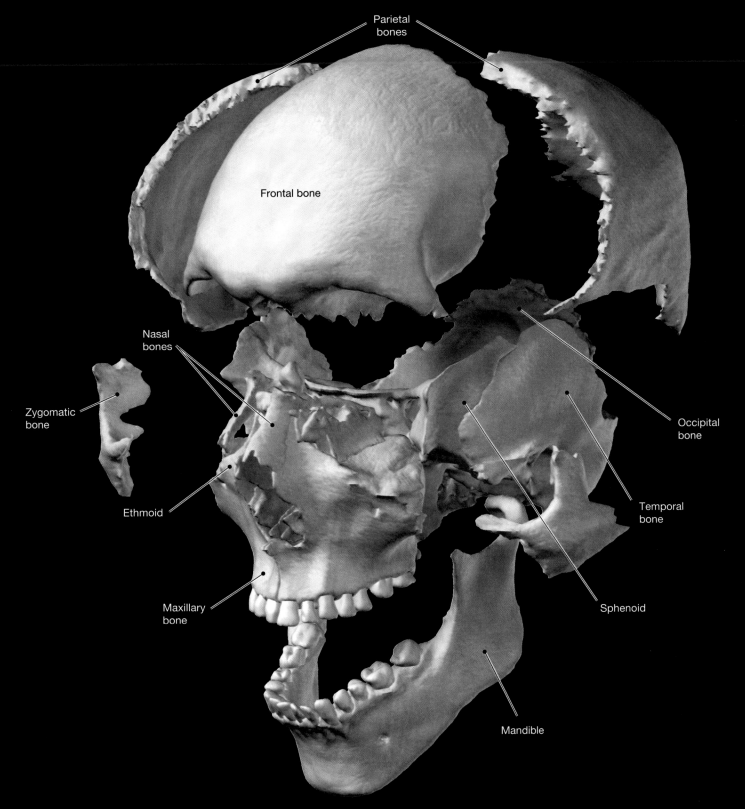

Parietal
bones

Frontal bone

Nasal
bones

Zygomatic
bone

Occipital
bone

Ethmoid

Temporal
bone

Maxillary
bone

Sphenoid

Mandible

PLATE **6** A COMPUTER RECONSTRUCTION OF A DISARTICULATED SKULL

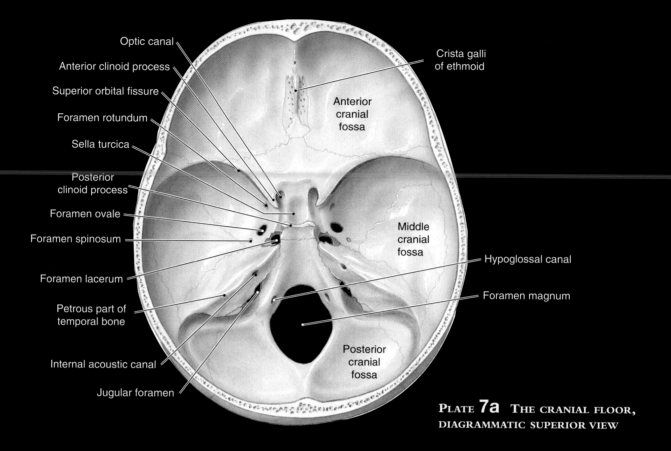

Optic canal

Anterior clinoid process

Superior orbital fissure

Foramen rotundum

Sella turcica

Posterior clinoid process

Foramen ovale

Foramen spinosum

Foramen lacerum

Petrous part of temporal bone

Internal acoustic canal

Jugular foramen

Crista galli of ethmoid

Anterior cranial fossa

Middle cranial fossa

Hypoglossal canal

Foramen magnum

Posterior cranial fossa

PLATE **7a** THE CRANIAL FLOOR, DIAGRAMMATIC SUPERIOR VIEW

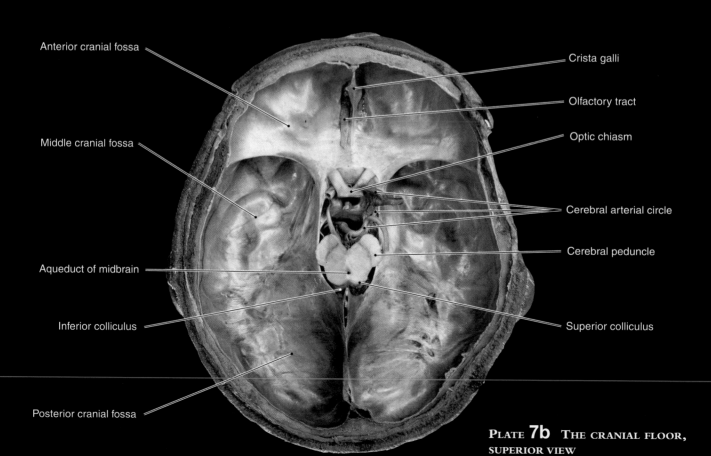

Anterior cranial fossa

Middle cranial fossa

Aqueduct of midbrain

Inferior colliculus

Posterior cranial fossa

Crista galli

Olfactory tract

Optic chiasm

Cerebral arterial circle

Cerebral peduncle

Superior colliculus

PLATE **7b** THE CRANIAL FLOOR, SUPERIOR VIEW

PLATE 7c **THE CRANIAL MENINGES, SUPERIOR VIEW**

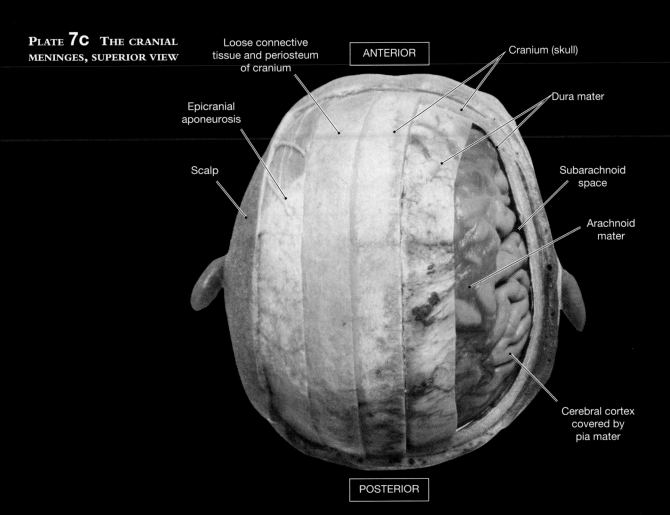

Loose connective tissue and periosteum of cranium

ANTERIOR

Cranium (skull)

Dura mater

Epicranial aponeurosis

Scalp

Subarachnoid space

Arachnoid mater

Cerebral cortex covered by pia mater

POSTERIOR

PLATE 7d **DURAL FOLDS**

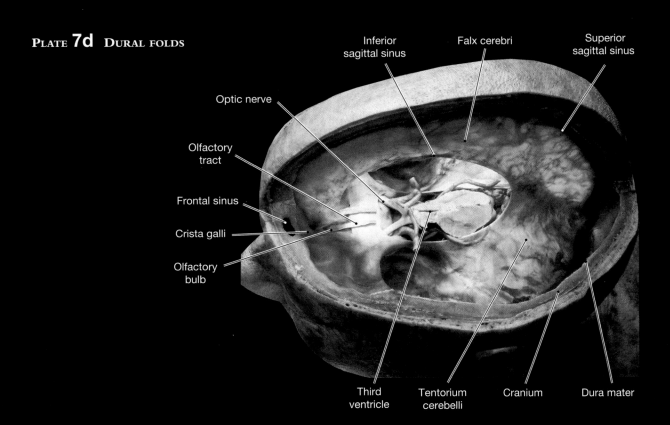

Inferior sagittal sinus

Falx cerebri

Superior sagittal sinus

Optic nerve

Olfactory tract

Frontal sinus

Crista galli

Olfactory bulb

Third ventricle

Tentorium cerebelli

Cranium

Dura mater

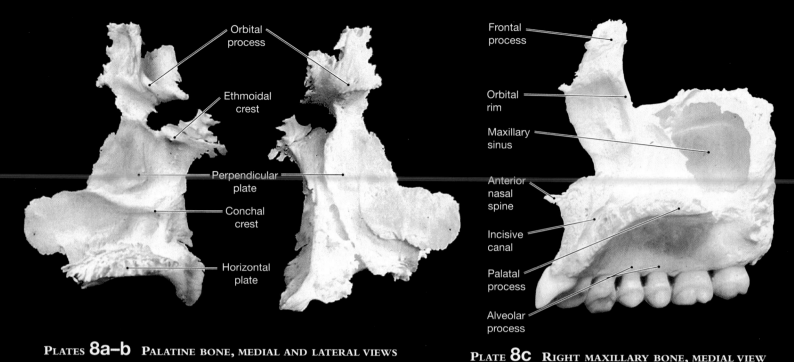

Orbital process

Ethmoidal crest

Perpendicular plate

Conchal crest

Horizontal plate

Frontal process

Orbital rim

Maxillary sinus

Anterior nasal spine

Incisive canal

Palatal process

Alveolar process

PLATES **8a–b** PALATINE BONE, MEDIAL AND LATERAL VIEWS

PLATE **8c** RIGHT MAXILLARY BONE, MEDIAL VIEW

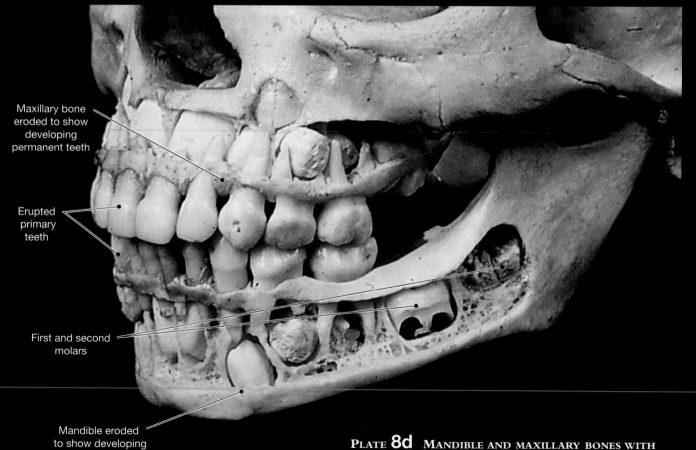

Maxillary bone eroded to show developing permanent teeth

Erupted primary teeth

First and second molars

Mandible eroded to show developing permanent teeth

PLATE **8d** MANDIBLE AND MAXILLARY BONES WITH UNERUPTED TEETH EXPOSED

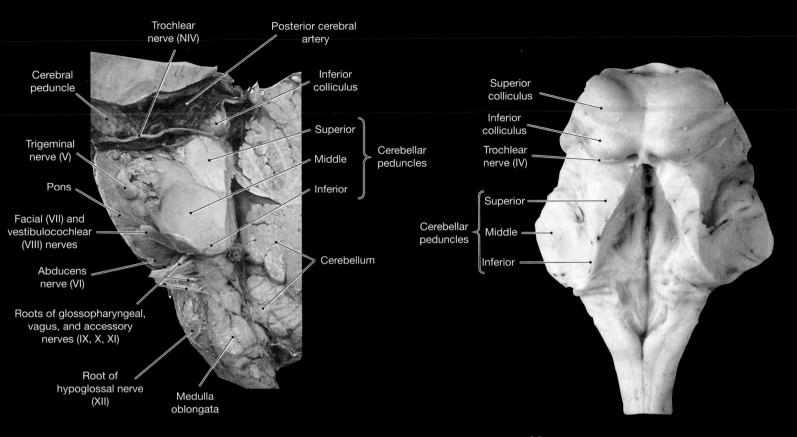

Trochlear nerve (NIV)

Posterior cerebral artery

Cerebral peduncle

Inferior colliculus

Trigeminal nerve (V)

Superior

Middle

Cerebellar peduncles

Pons

Inferior

Facial (VII) and vestibulocochlear (VIII) nerves

Abducens nerve (VI)

Cerebellum

Roots of glossopharyngeal, vagus, and accessory nerves (IX, X, XI)

Root of hypoglossal nerve (XII)

Medulla oblongata

Superior colliculus

Inferior colliculus

Trochlear nerve (IV)

Superior

Cerebellar peduncles

Middle

Inferior

PLATE 9a BRAIN STEM, LATERAL VIEW

PLATE 9b BRAIN STEM, POSTERIOR VIEW

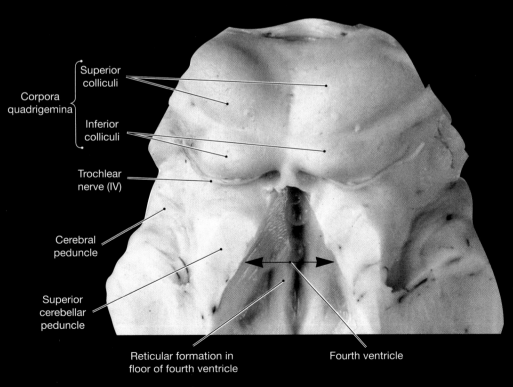

Superior colliculi

Corpora quadrigemina

Inferior colliculi

Trochlear nerve (IV)

Cerebral peduncle

Superior cerebellar peduncle

Reticular formation in floor of fourth ventricle

Fourth ventricle

PLATE 9c THE MESENCEPHALON, POSTERIOR VIEW

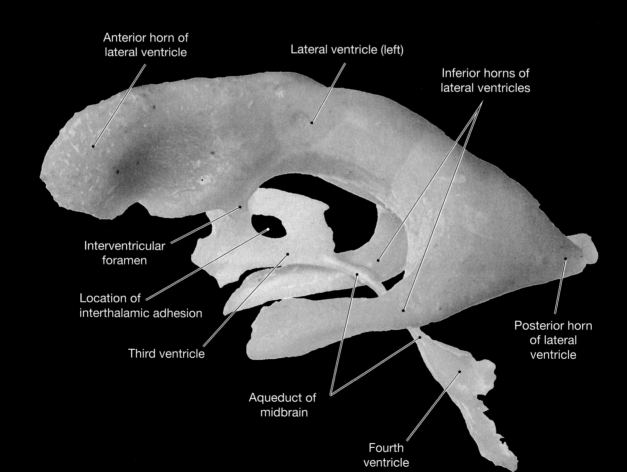

Anterior horn of
lateral ventricle

Lateral ventricle (left)

Inferior horns of
lateral ventricles

Interventricular
foramen

Location of
interthalamic adhesion

Third ventricle

Posterior horn
of lateral
ventricle

Aqueduct of
midbrain

Fourth
ventricle

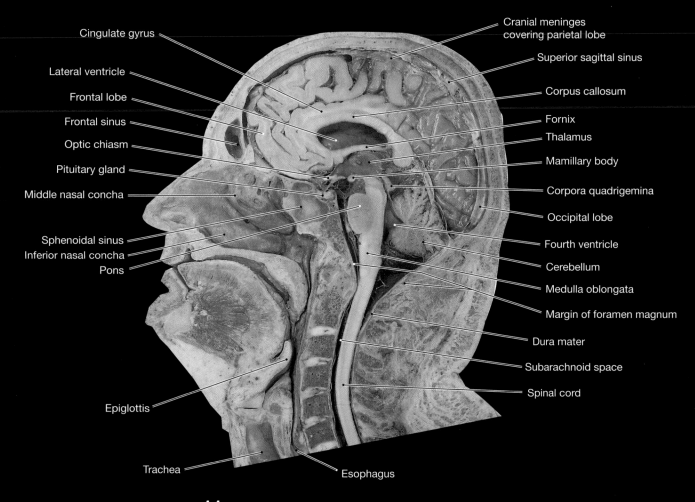

Cingulate gyrus

Lateral ventricle

Frontal lobe

Frontal sinus

Optic chiasm

Pituitary gland

Middle nasal concha

Sphenoidal sinus

Inferior nasal concha

Pons

Epiglottis

Trachea

Esophagus

Cranial meninges covering parietal lobe

Superior sagittal sinus

Corpus callosum

Fornix

Thalamus

Mamillary body

Corpora quadrigemina

Occipital lobe

Fourth ventricle

Cerebellum

Medulla oblongata

Margin of foramen magnum

Dura mater

Subarachnoid space

Spinal cord

PLATE **11a** MIDSAGITTAL SECTION THROUGH THE HEAD AND NECK

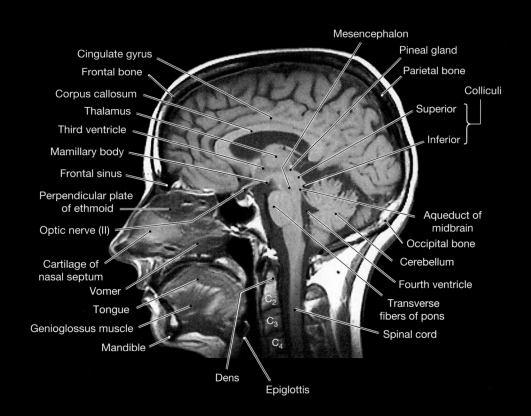

Cingulate gyrus

Frontal bone

Corpus callosum

Thalamus

Third ventricle

Mamillary body

Frontal sinus

Perpendicular plate of ethmoid

Optic nerve (II)

Cartilage of nasal septum

Vomer

Tongue

Genioglossus muscle

Mandible

Dens

Epiglottis

Mesencephalon

Pineal gland

Parietal bone

Colliculi

Superior

Inferior

Aqueduct of midbrain

Occipital bone

Cerebellum

Fourth ventricle

Transverse fibers of pons

Spinal cord

C_2

C_3

C_4

PLATE **11b** MRI SCAN OF THE BRAIN, MIDSAGITTAL SECTION

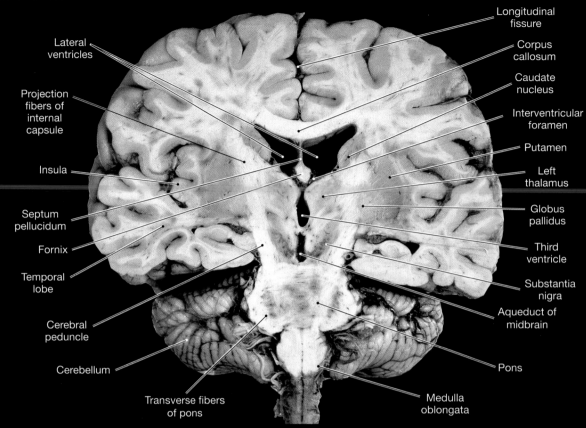

Longitudinal fissure

Corpus callosum

Caudate nucleus

Interventricular foramen

Putamen

Left thalamus

Globus pallidus

Third ventricle

Substantia nigra

Aqueduct of midbrain

Pons

Medulla oblongata

Lateral ventricles

Projection fibers of internal capsule

Insula

Septum pellucidum

Fornix

Temporal lobe

Cerebral peduncle

Cerebellum

Transverse fibers of pons

PLATE **11c** CORONAL SECTION THROUGH THE BRAIN AT THE LEVEL OF THE MESENCEPHALON AND PONS

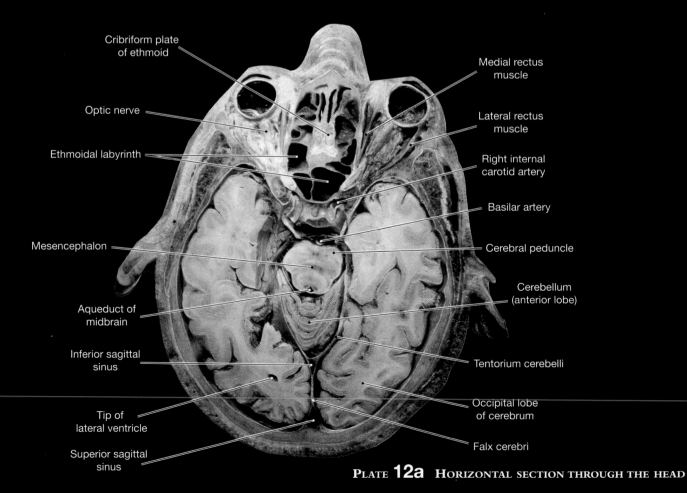

Cribriform plate of ethmoid

Medial rectus muscle

Optic nerve

Lateral rectus muscle

Ethmoidal labyrinth

Right internal carotid artery

Basilar artery

Mesencephalon

Cerebral peduncle

Cerebellum (anterior lobe)

Aqueduct of midbrain

Inferior sagittal sinus

Tentorium cerebelli

Occipital lobe of cerebrum

Tip of lateral ventricle

Falx cerebri

Superior sagittal sinus

PLATE **12a** HORIZONTAL SECTION THROUGH THE HEAD

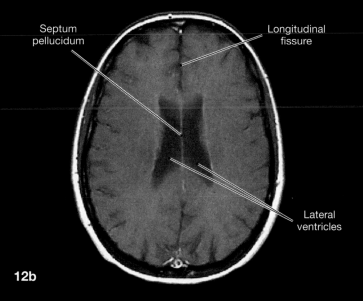

Septum pellucidum

Longitudinal fissure

Lateral ventricles

12b

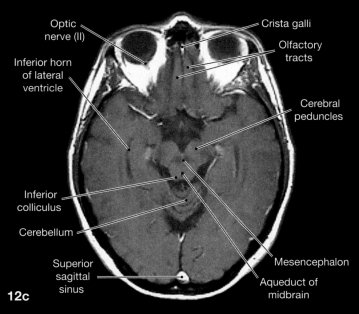

Optic nerve (II)

Inferior horn of lateral ventricle

Inferior colliculus

Cerebellum

Superior sagittal sinus

Crista galli

Olfactory tracts

Cerebral peduncles

Mesencephalon

Aqueduct of midbrain

12c

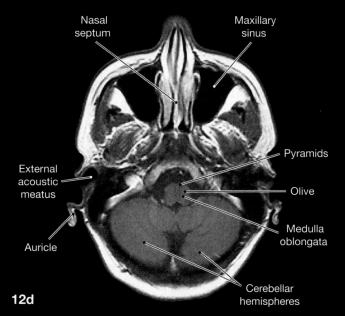

Nasal septum

Maxillary sinus

External acoustic meatus

Auricle

Pyramids

Olive

Medulla oblongata

Cerebellar hemispheres

12d

PLATES **12b–d** MRI SCANS OF THE BRAIN, HORIZONTAL SECTIONS, SUPERIOR TO INFERIOR SEQUENCE

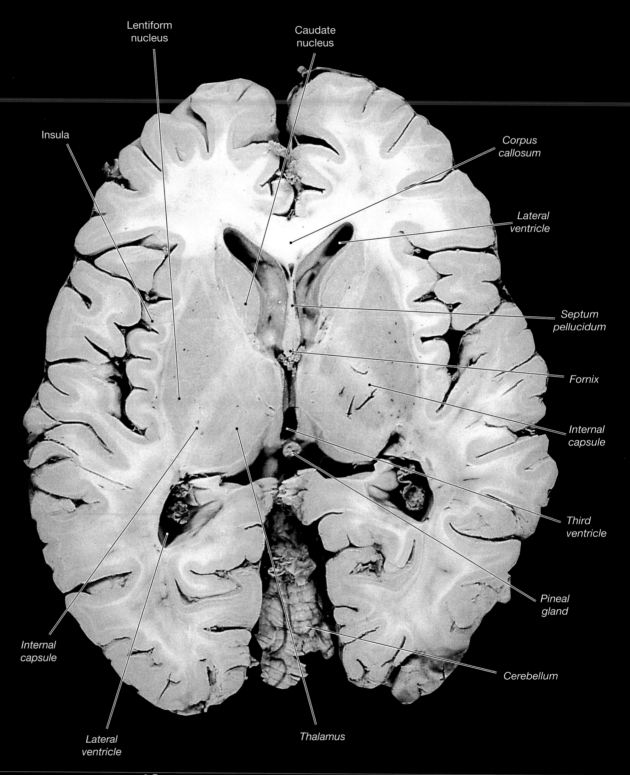

Lentiform
nucleus

Caudate
nucleus

Insula

*Corpus
callosum*

*Lateral
ventricle*

*Septum
pellucidum*

Fornix

*Internal
capsule*

*Third
ventricle*

*Pineal
gland*

*Internal
capsule*

Cerebellum

*Lateral
ventricle*

Thalamus

PLATE **13a** A HORIZONTAL SECTION THROUGH THE BRAIN

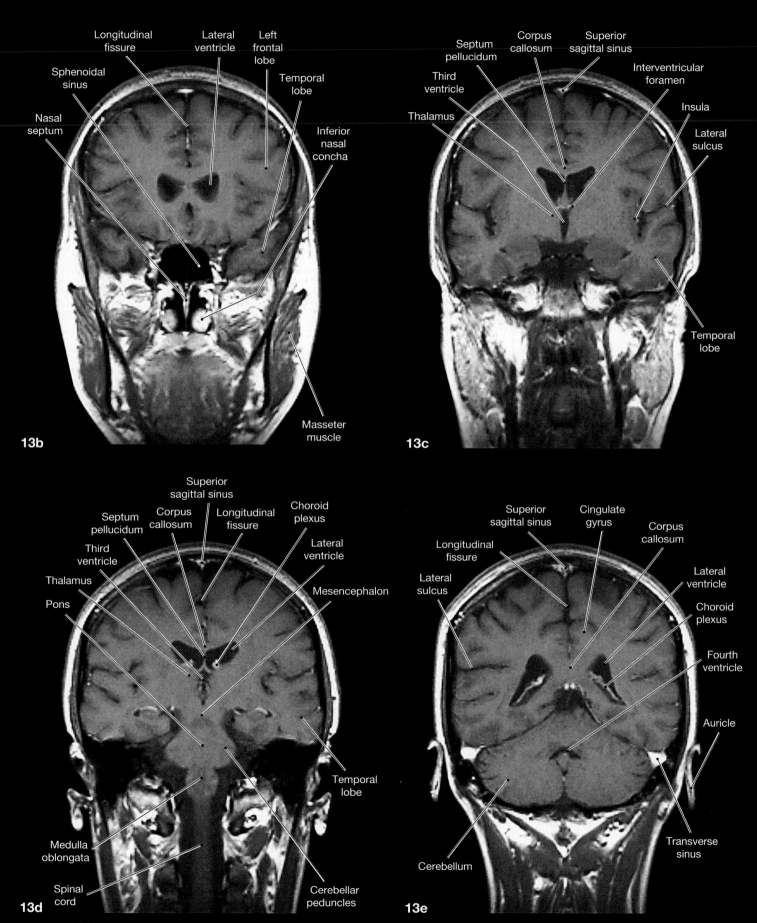

13b

Longitudinal fissure
Lateral ventricle
Left frontal lobe
Sphenoidal sinus
Temporal lobe
Nasal septum
Inferior nasal concha
Masseter muscle

13c

Septum pellucidum
Corpus callosum
Superior sagittal sinus
Third ventricle
Interventricular foramen
Thalamus
Insula
Lateral sulcus
Temporal lobe

13d

Superior sagittal sinus
Septum pellucidum
Corpus callosum
Longitudinal fissure
Choroid plexus
Lateral ventricle
Third ventricle
Mesencephalon
Thalamus
Pons
Temporal lobe
Medulla oblongata
Spinal cord
Cerebellar peduncles

13e

Superior sagittal sinus
Cingulate gyrus
Corpus callosum
Longitudinal fissure
Lateral sulcus
Lateral ventricle
Choroid plexus
Fourth ventricle
Auricle
Transverse sinus
Cerebellum

PLATES **13b–e** MRI SCANS OF THE BRAIN, FRONTAL (CORONAL) SECTIONS, ANTERIOR TO POSTERIOR SEQUENCE

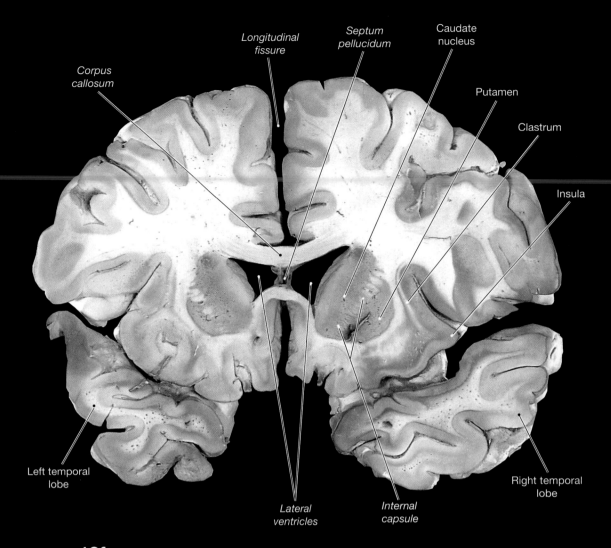

Corpus callosum

Longitudinal fissure

Septum pellucidum

Caudate nucleus

Putamen

Clastrum

Insula

Left temporal lobe

Right temporal lobe

Lateral ventricles

Internal capsule

PLATE **13f** FRONTAL SECTION THROUGH THE BRAIN AT THE LEVEL OF THE BASAL NUCLEI

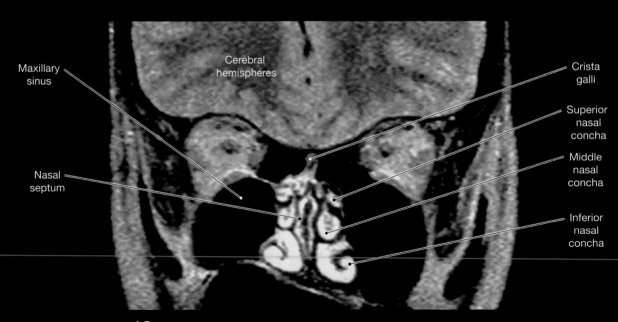

Maxillary sinus

Cerebral hemispheres

Crista galli

Superior nasal concha

Middle nasal concha

Nasal septum

Inferior nasal concha

PLATE **13g** MRI SCAN, CORONAL SECTION SHOWING PARANASAL SINUSES

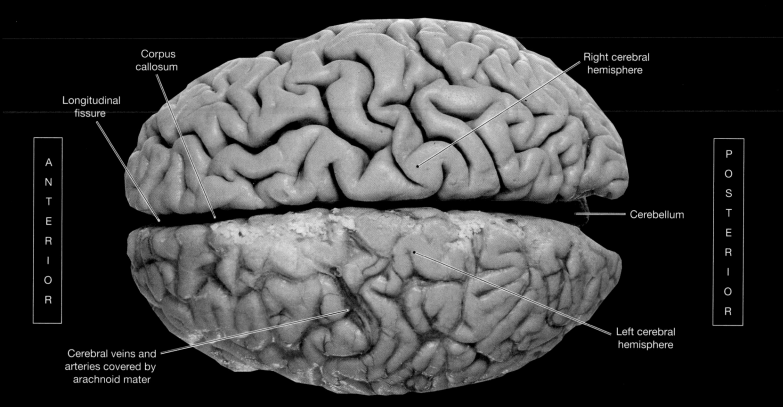

Corpus callosum

Longitudinal fissure

Right cerebral hemisphere

ANTERIOR

POSTERIOR

Cerebellum

Left cerebral hemisphere

Cerebral veins and arteries covered by arachnoid mater

PLATE **14a** SURFACE ANATOMY OF THE BRAIN, SUPERIOR VIEW

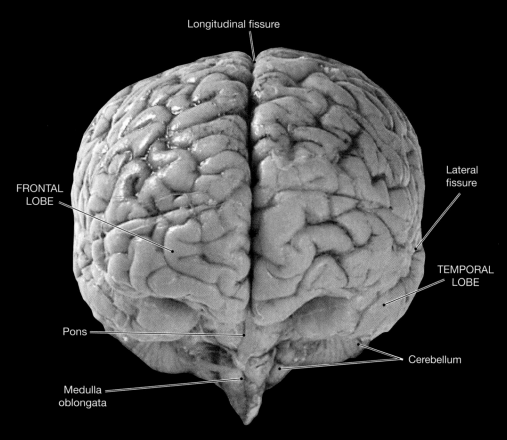

Longitudinal fissure

Lateral fissure

FRONTAL LOBE

TEMPORAL LOBE

Pons

Cerebellum

Medulla oblongata

PLATE **14b** SURFACE ANATOMY OF THE BRAIN, ANTERIOR VIEW

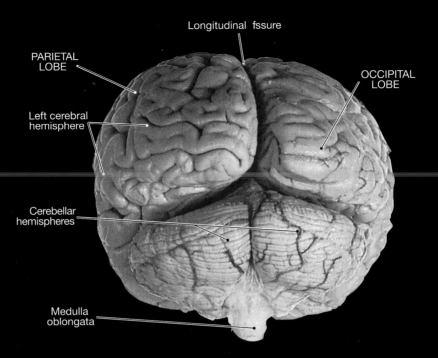

Longitudinal fssure

PARIETAL
LOBE

OCCIPITAL
LOBE

Left cerebral
hemisphere

Cerebellar
hemispheres

Medulla
oblongata

PLATE **14c** SURFACE ANATOMY OF THE BRAIN, POSTERIOR VIEW

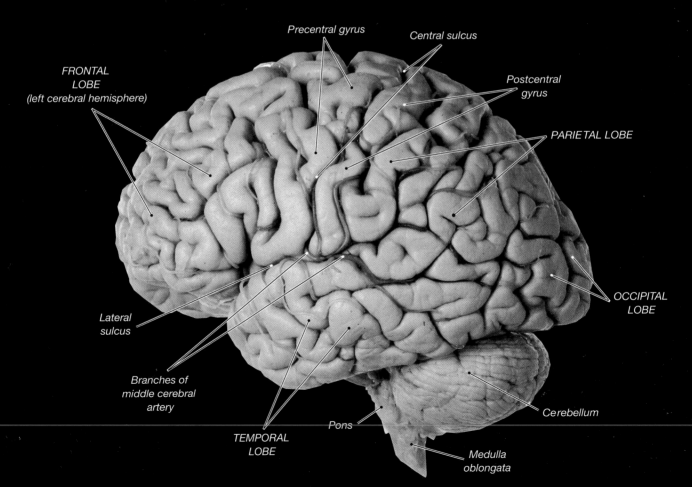

Precentral gyrus

Central sulcus

FRONTAL
LOBE
(left cerebral hemisphere)

Postcentral
gyrus

PARIETAL LOBE

OCCIPITAL
LOBE

Lateral
sulcus

Branches of
middle cerebral
artery

Cerebellum

Pons

TEMPORAL
LOBE

Medulla
oblongata

PLATE **14d** SURFACE ANATOMY OF THE BRAIN, LATERAL VIEW

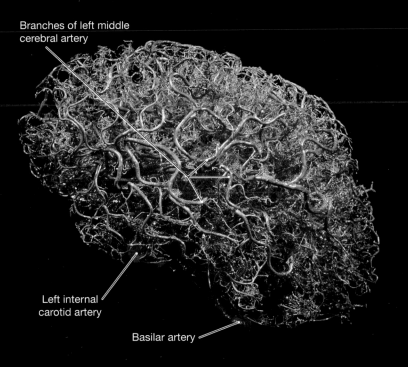

Branches of left middle cerebral artery

Left internal carotid artery

Basilar artery

PLATE 15a ARTERIAL CIRCULATION TO THE BRAIN, LATERAL VIEW OF CORROSION CAST

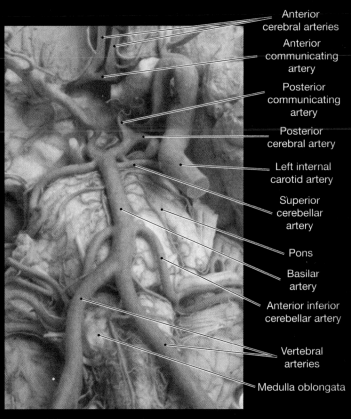

Anterior cerebral arteries

Anterior communicating artery

Posterior communicating artery

Posterior cerebral artery

Left internal carotid artery

Superior cerebellar artery

Pons

Basilar artery

Anterior inferior cerebellar artery

Vertebral arteries

Medulla oblongata

PLATE 15c ARTERIES ON THE INFERIOR SURFACE OF THE BRAIN

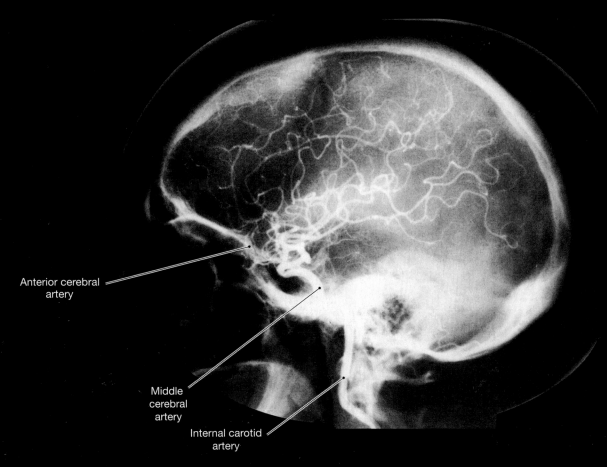

Anterior cerebral artery

Middle cerebral artery

Internal carotid artery

PLATE 15b CRANIAL ANGIOGRAM

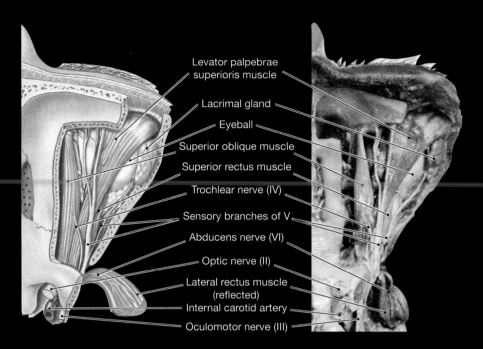

Levator palpebrae superioris muscle

Lacrimal gland

Eyeball

Superior oblique muscle

Superior rectus muscle

Trochlear nerve (IV)

Sensory branches of V

Abducens nerve (VI)

Optic nerve (II)

Lateral rectus muscle (reflected)

Internal carotid artery

Oculomotor nerve (III)

PLATES **16a–b** ACCESSORY STRUCTURES OF THE EYE, SUPERIOR VIEW

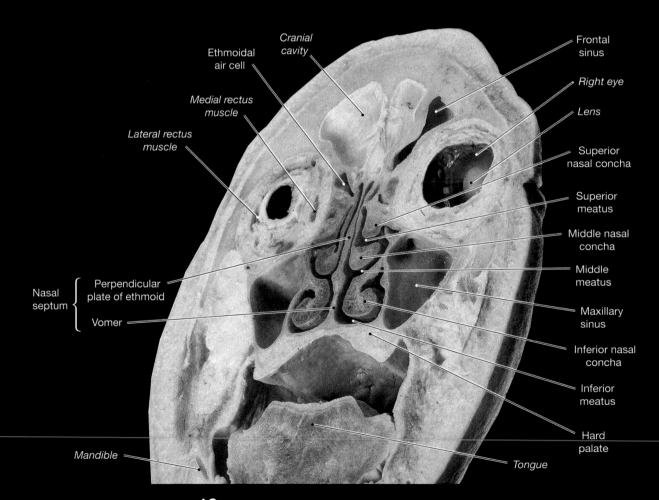

Cranial cavity

Ethmoidal air cell

Medial rectus muscle

Lateral rectus muscle

Nasal septum
  Perpendicular plate of ethmoid
  Vomer

Mandible

Frontal sinus

Right eye

Lens

Superior nasal concha

Superior meatus

Middle nasal concha

Middle meatus

Maxillary sinus

Inferior nasal concha

Inferior meatus

Hard palate

Tongue

PLATE **16c** FRONTAL SECTION THROUGH THE FACE

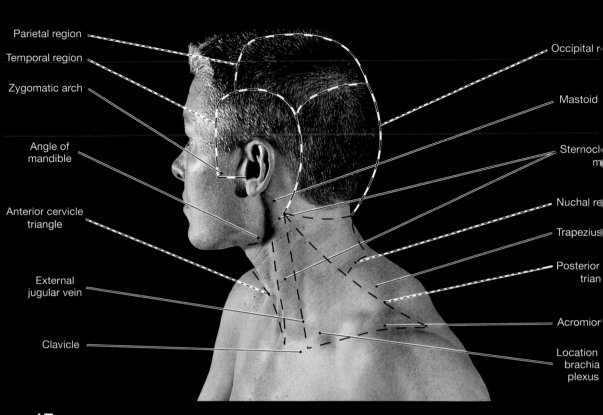

Parietal region

Temporal region

Zygomatic arch

Angle of
mandible

Anterior cervicle
triangle

External
jugular vein

Clavicle

Occipital r

Mastoid

Sternocl
m

Nuchal re

Trapezius

Posterior
trian

Acromior

Location
brachia
plexus

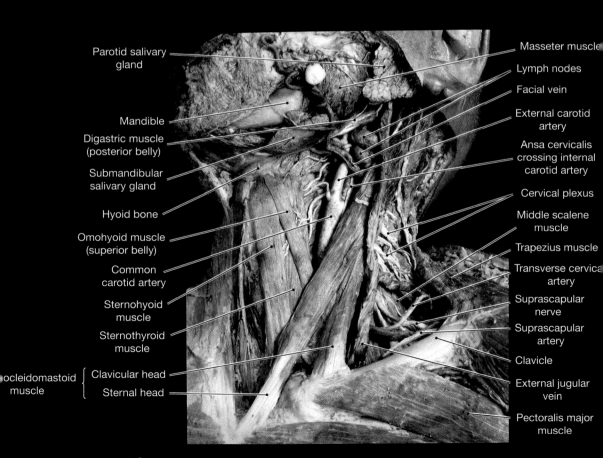

Parotid salivary
gland

Mandible

Digastric muscle
(posterior belly)

Submandibular
salivary gland

Hyoid bone

Omohyoid muscle
(superior belly)

Common
carotid artery

Sternohyoid
muscle

Sternothyroid
muscle

ocleidomastoid
muscle { Clavicular head
Sternal head

Masseter muscle

Lymph nodes

Facial vein

External carotid
artery

Ansa cervicalis
crossing internal
carotid artery

Cervical plexus

Middle scalene
muscle

Trapezius muscle

Transverse cervica
artery

Suprascapular
nerve

Suprascapular
artery

Clavicle

External jugular
vein

Pectoralis major
muscle

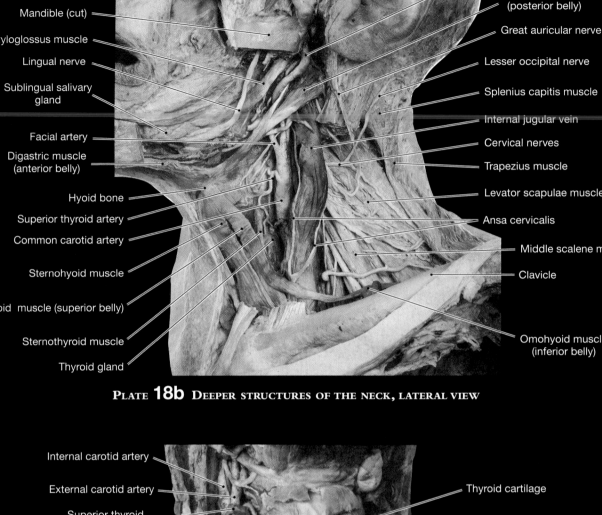

Zygomatic arch
Temporalis muscle
Mandible (cut)
yloglossus muscle
Lingual nerve
Sublingual salivary gland
Facial artery
Digastric muscle (anterior belly)
Hyoid bone
Superior thyroid artery
Common carotid artery
Sternohyoid muscle
oid muscle (superior belly)
Sternothyroid muscle
Thyroid gland

External carotid arter
Digastric muscle (posterior belly)
Great auricular nerve
Lesser occipital nerve
Splenius capitis muscle
Internal jugular vein
Cervical nerves
Trapezius muscle
Levator scapulae muscle
Ansa cervicalis
Middle scalene m
Clavicle
Omohyoid muscl (inferior belly)

PLATE 18b DEEPER STRUCTURES OF THE NECK, LATERAL VIEW

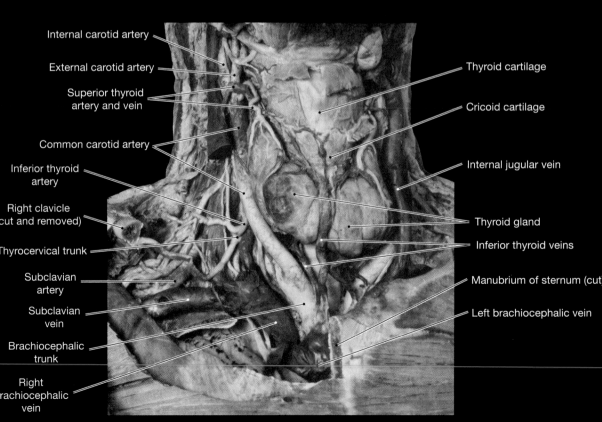

Internal carotid artery
External carotid artery
Superior thyroid artery and vein
Common carotid artery
Inferior thyroid artery
Right clavicle (cut and removed)
Thyrocervical trunk
Subclavian artery
Subclavian vein
Brachiocephalic trunk
Right brachiocephalic vein

Thyroid cartilage
Cricoid cartilage
Internal jugular vein
Thyroid gland
Inferior thyroid veins
Manubrium of sternum (cut)
Left brachiocephalic vein

PLATE 18c DEEPER STRUCTURES OF THE NECK, ANTERIOR VIEW

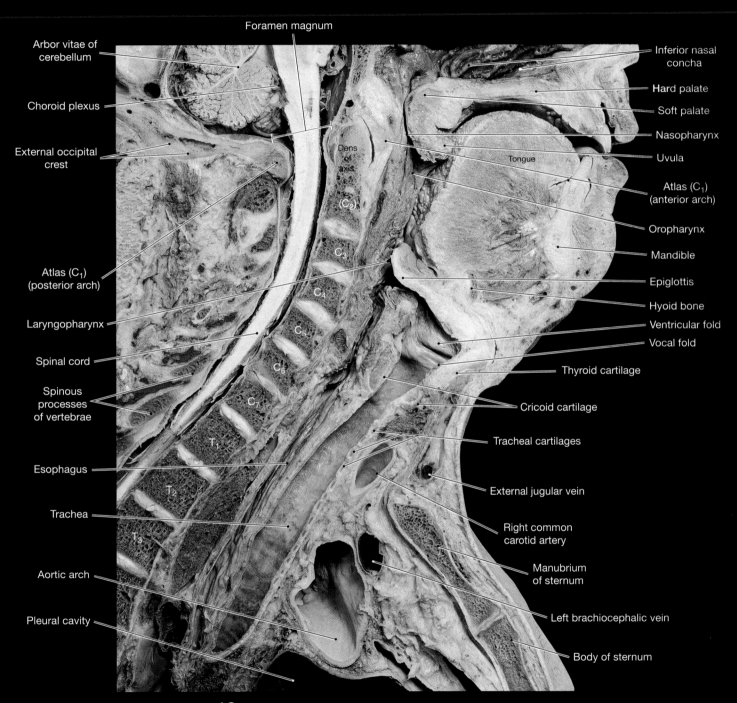

Foramen magnum

Arbor vitae of cerebellum

Choroid plexus

External occipital crest

Atlas (C₁) (posterior arch)

Laryngopharynx

Spinal cord

Spinous processes of vertebrae

Esophagus

Trachea

Aortic arch

Pleural cavity

Dens of axis

(C₂)

C₃

C₄

C₅

C₆

C₇

T₁

T₂

T₃

Inferior nasal concha

Hard palate

Soft palate

Nasopharynx

Uvula

Atlas (C₁) (anterior arch)

Oropharynx

Mandible

Epiglottis

Hyoid bone

Ventricular fold

Vocal fold

Thyroid cartilage

Cricoid cartilage

Tracheal cartilages

External jugular vein

Right common carotid artery

Manubrium of sternum

Left brachiocephalic vein

Body of sternum

Tongue

PLATE 19 MIDSAGITAL SECTION THROUGH THE HEAD AND NECK

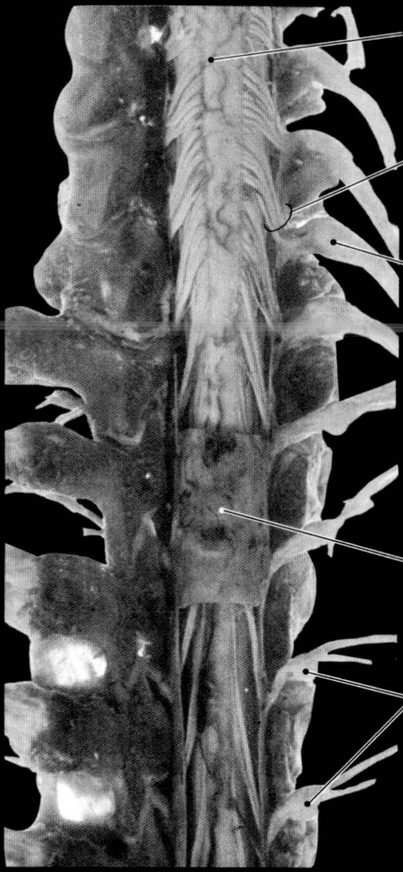

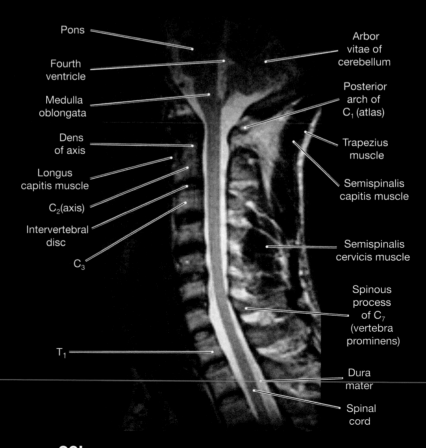

Cervical
spinal cord

Rootlets
of C_8

Dorsal root
ganglion of C_8

PLATE 20a THE CERVICAL AND
THORACIC REGIONS OF THE SPINAL
CORD, POSTERIOR VIEW

Dura mater

Dorsal root
ganglia of T_4
and T_5

Pons

Arbor
vitae of
cerebellum

Fourth
ventricle

Posterior
arch of
C_1 (atlas)

Medulla
oblongata

Dens
of axis

Trapezius
muscle

Longus
capitis muscle

Semispinalis
capitis muscle

C_2(axis)

Intervertebral
disc

C_3

Semispinalis
cervicis muscle

Spinous
process
of C_7
(vertebra
prominens)

T_1

Dura
mater

Spinal
cord

PLATE 20b MRI SCAN OF CERVICAL REGION, SAGITTAL SECTION

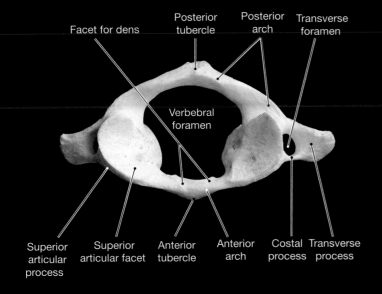

Facet for dens — Posterior tubercle — Posterior arch — Transverse foramen

Verbebral foramen

Superior articular process — Superior articular facet — Anterior tubercle — Anterior arch — Costal process — Transverse process

PLATE 21a ATLAS, SUPERIOR VIEW

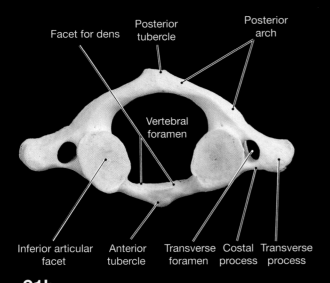

Facet for dens — Posterior tubercle — Posterior arch

Vertebral foramen

Inferior articular facet — Anterior tubercle — Transverse foramen — Costal process — Transverse process

PLATE 21b ATLAS, INFERIOR VIEW

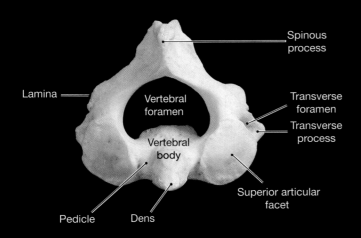

Lamina — Spinous process

Vertebral foramen — Transverse foramen — Transverse process

Vertebral body

Pedicle — Dens — Superior articular facet

PLATE 21c AXIS, SUPERIOR VIEW

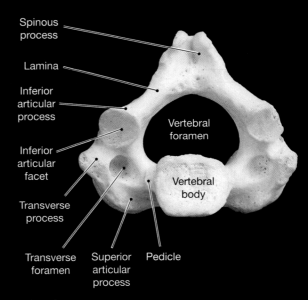

Spinous process — Lamina — Inferior articular process — Inferior articular facet — Transverse process

Vertebral foramen

Vertebral body

Transverse foramen — Superior articular process — Pedicle

PLATE 21d AXIS, INFERIOR VIEW

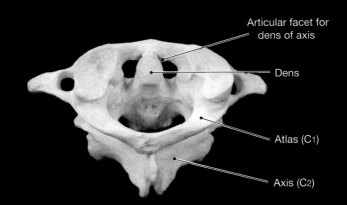

Articular facet for dens of axis — Dens

Atlas (C1)

Axis (C2)

PLATE 21e ARTICULATED ATLAS AND AXIS, SUPERIOR AND POSTERIOR VIEW

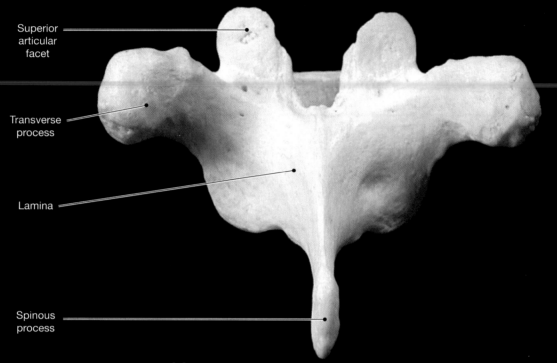

Superior
articular
facet

Transverse
process

Lamina

Spinous
process

PLATE **22a**   THORACIC VERTEBRA, POSTERIOR VIEW

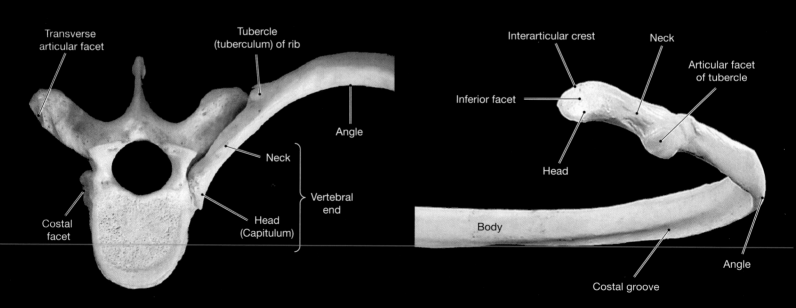

Transverse
articular facet

Tubercle
(tuberculum) of rib

Angle

Neck

Vertebral
end

Head
(Capitulum)

Costal
facet

PLATE **22b**   THORACIC VERTEBRA AND RIB, SUPERIOR VIEW

Interarticular crest

Neck

Articular facet
of tubercle

Inferior facet

Head

Body

Costal groove

Angle

PLATE **22c**   REPRESENTATIVE RIB, POSTERIOR VIEW

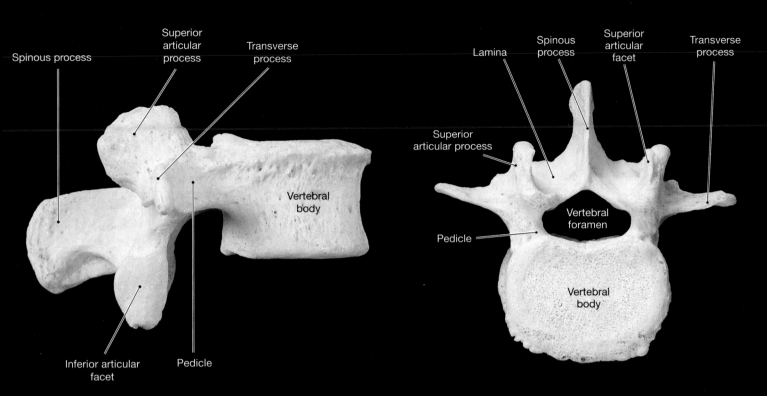

Spinous process

Superior articular process

Transverse process

Vertebral body

Inferior articular facet

Pedicle

PLATE 23a THE LUMBAR VERTEBRAE, LATERAL VIEW

Lamina

Spinous process

Superior articular facet

Transverse process

Superior articular process

Vertebral foramen

Pedicle

Vertebral body

PLATE 23b THE LUMBAR VERTEBRAE, SUPERIOR VIEW

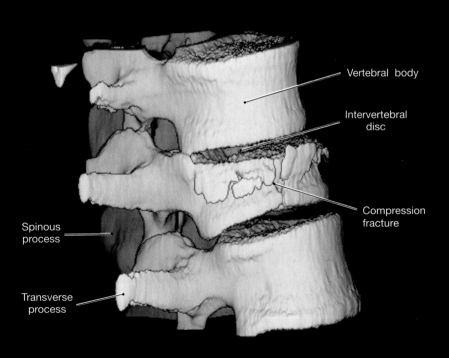

Vertebral body

Intervertebral disc

Compression fracture

Spinous process

Transverse process

PLATE 23c 3-DIMENSIONAL CT SCAN SHOWING A FRACTURE OF THE BODY OF A LUMBAR VERTEBRA

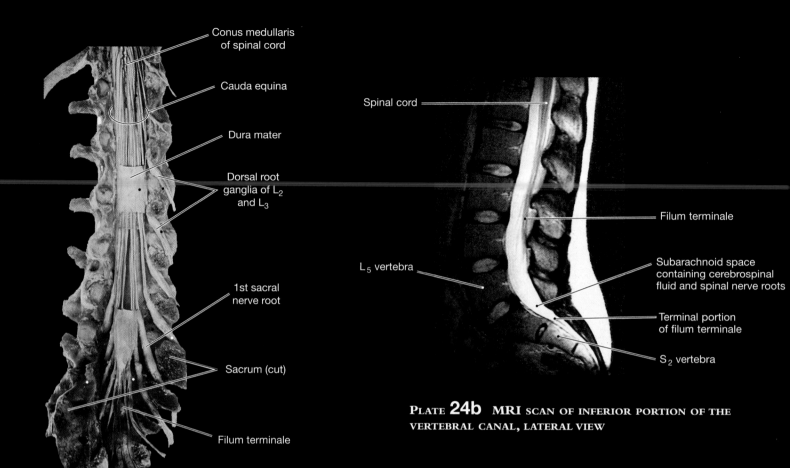

Conus medullaris of spinal cord

Cauda equina

Dura mater

Dorsal root ganglia of L_2 and L_3

1st sacral nerve root

Sacrum (cut)

Filum terminale

PLATE 24a THE INFERIOR PORTION OF THE VERTEBRAL COLUMN, POSTERIOR VIEW

Spinal cord

Filum terminale

L_5 vertebra

Subarachnoid space containing cerebrospinal fluid and spinal nerve roots

Terminal portion of filum terminale

S_2 vertebra

PLATE 24b MRI SCAN OF INFERIOR PORTION OF THE VERTEBRAL CANAL, LATERAL VIEW

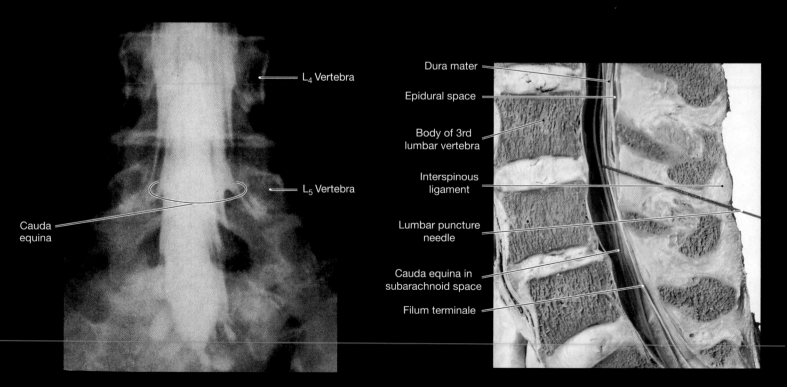

L_4 Vertebra

L_5 Vertebra

Cauda equina

PLATE 24c X-RAY OF THE CAUDA EQUINA WITH A CONTRAST MEDIUM IN THE SUBARACHNOID SPACE

Dura mater

Epidural space

Body of 3rd lumbar vertebra

Interspinous ligament

Lumbar puncture needle

Cauda equina in subarachnoid space

Filum terminale

PLATE 24d LUMBAR PUNCTURE POSITIONING, SEEN IN A SAGITTAL SECTION

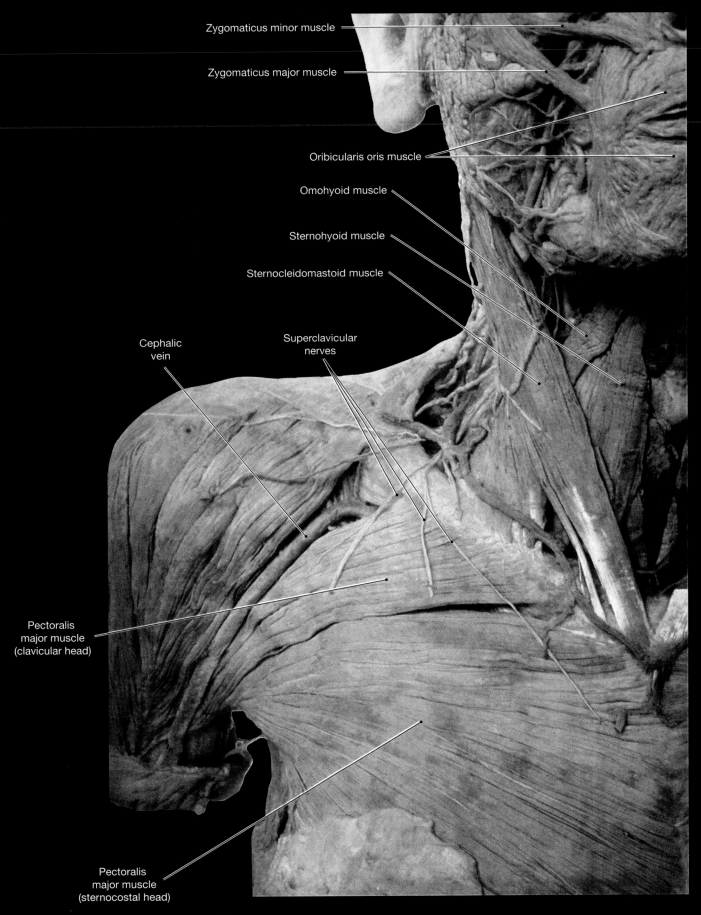

Zygomaticus minor muscle

Zygomaticus major muscle

Oribicularis oris muscle

Omohyoid muscle

Sternohyoid muscle

Sternocleidomastoid muscle

Cephalic
vein

Superclavicular
nerves

Pectoralis
major muscle
(clavicular head)

Pectoralis
major muscle
(sternocostal head)

PLATE 25   SHOULDER AND NECK, ANTERIOR VIEW

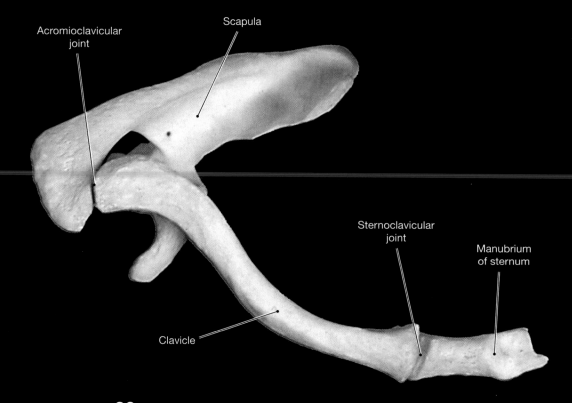

Acromioclavicular
joint

Scapula

Sternoclavicular
joint

Manubrium
of sternum

Clavicle

PLATE **26a** BONES OF THE RIGHT PECTORAL GIRDLE, SUPERIOR VIEW

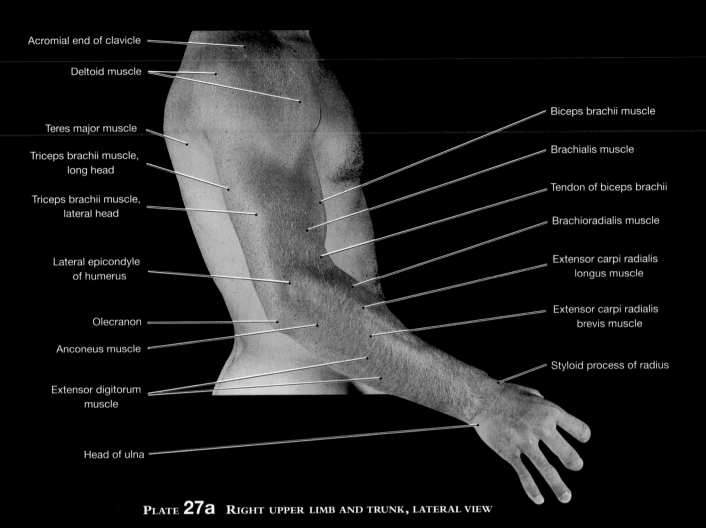

Acromial end of clavicle

Deltoid muscle

Teres major muscle

Triceps brachii muscle, long head

Triceps brachii muscle, lateral head

Lateral epicondyle of humerus

Olecranon

Anconeus muscle

Extensor digitorum muscle

Head of ulna

Biceps brachii muscle

Brachialis muscle

Tendon of biceps brachii

Brachioradialis muscle

Extensor carpi radialis longus muscle

Extensor carpi radialis brevis muscle

Styloid process of radius

PLATE **27a**   RIGHT UPPER LIMB AND TRUNK, LATERAL VIEW

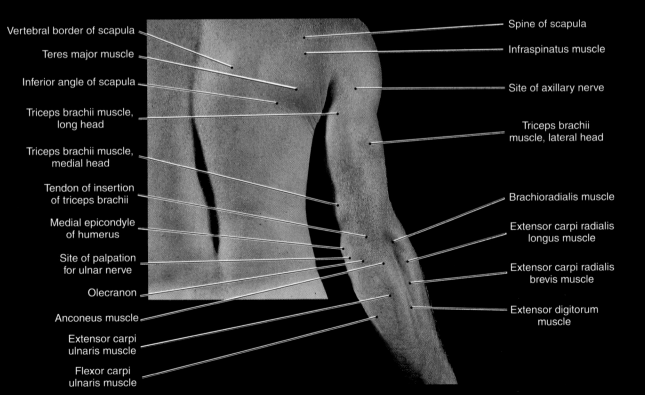

Vertebral border of scapula

Teres major muscle

Inferior angle of scapula

Triceps brachii muscle, long head

Triceps brachii muscle, medial head

Tendon of insertion of triceps brachii

Medial epicondyle of humerus

Site of palpation for ulnar nerve

Olecranon

Anconeus muscle

Extensor carpi ulnaris muscle

Flexor carpi ulnaris muscle

Spine of scapula

Infraspinatus muscle

Site of axillary nerve

Triceps brachii muscle, lateral head

Brachioradialis muscle

Extensor carpi radialis longus muscle

Extensor carpi radialis brevis muscle

Extensor digitorum muscle

PLATE **27b**   RIGHT UPPER LIMB AND TRUNK, POSTERIOR VIEW

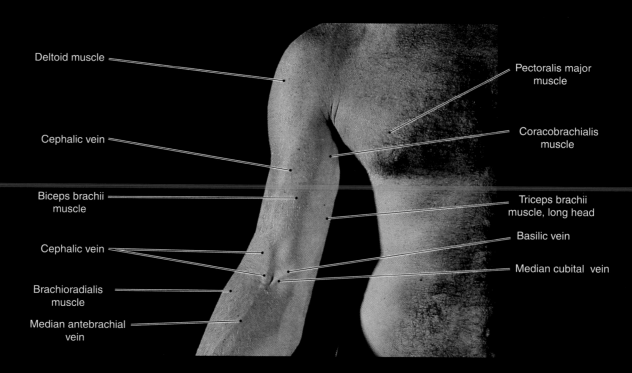

Deltoid muscle

Cephalic vein

Biceps brachii
muscle

Cephalic vein

Brachioradialis
muscle

Median antebrachial
vein

Pectoralis major
muscle

Coracobrachialis
muscle

Triceps brachii
muscle, long head

Basilic vein

Median cubital vein

PLATE **27c** RIGHT ARM AND TRUNK, ANTERIOR VIEW

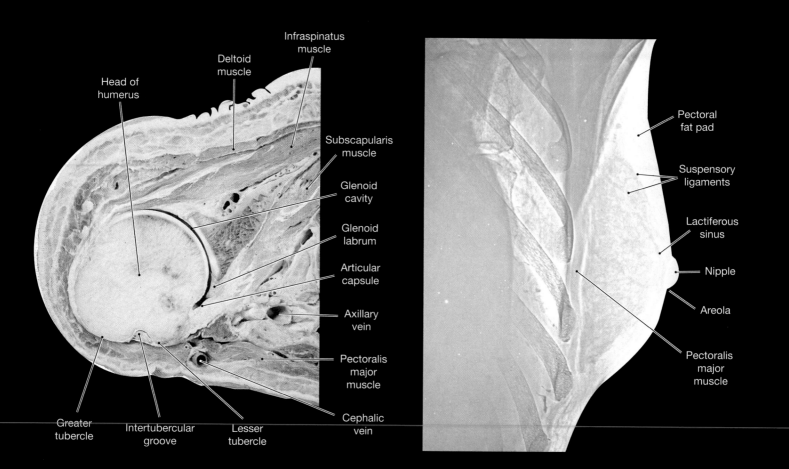

Infraspinatus
muscle

Deltoid
muscle

Head of
humerus

Subscapularis
muscle

Glenoid
cavity

Glenoid
labrum

Articular
capsule

Axillary
vein

Pectoralis
major
muscle

Cephalic
vein

Greater
tubercle

Intertubercular
groove

Lesser
tubercle

Pectoral
fat pad

Suspensory
ligaments

Lactiferous
sinus

Nipple

Areola

Pectoralis
major
muscle

PLATE **27d** HORIZONTAL SECTION THROUGH THE RIGHT
SHOULDER

PLATE **28** XEROMAMMOGRAM

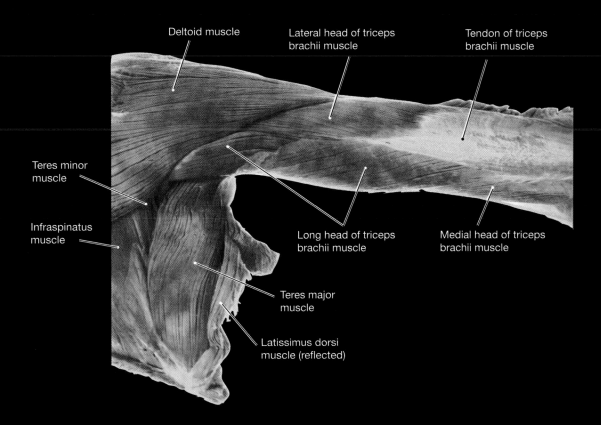

Deltoid muscle

Lateral head of triceps
brachii muscle

Tendon of triceps
brachii muscle

Teres minor
muscle

Infraspinatus
muscle

Long head of triceps
brachii muscle

Medial head of triceps
brachii muscle

Teres major
muscle

Latissimus dorsi
muscle (reflected)

PLATE **29a** SUPERFICIAL DISSECTION OF RIGHT SHOULDER, POSTERIOR VIEW

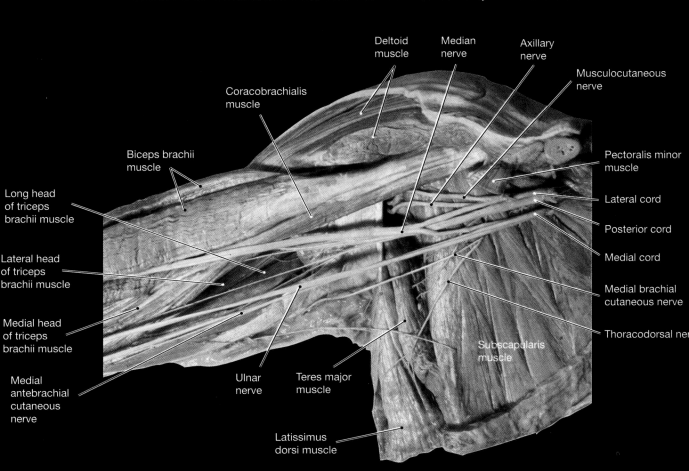

Coracobrachialis
muscle

Deltoid
muscle

Median
nerve

Axillary
nerve

Musculocutaneous
nerve

Biceps brachii
muscle

Pectoralis minor
muscle

Long head
of triceps
brachii muscle

Lateral cord

Posterior cord

Lateral head
of triceps
brachii muscle

Medial cord

Medial head
of triceps
brachii muscle

Medial brachial
cutaneous nerve

Thoracodorsal ner

Subscapularis
muscle

Medial
antebrachial
cutaneous
nerve

Ulnar
nerve

Teres major
muscle

Latissimus
dorsi muscle

PLATE **29b** DEEP DISSECTION OF THE RIGHT BRACHIAL PLEXUS

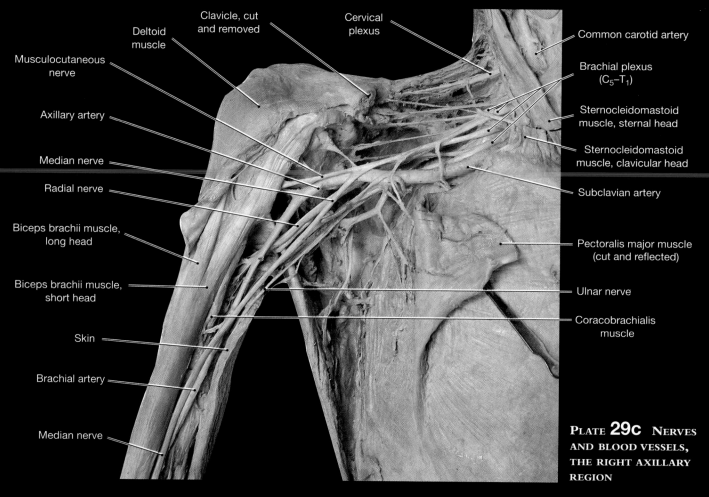

Clavicle, cut and removed

Cervical plexus

Deltoid muscle

Musculocutaneous nerve

Axillary artery

Median nerve

Radial nerve

Biceps brachii muscle, long head

Biceps brachii muscle, short head

Skin

Brachial artery

Median nerve

Common carotid artery

Brachial plexus (C₅–T₁)

Sternocleidomastoid muscle, sternal head

Sternocleidomastoid muscle, clavicular head

Subclavian artery

Pectoralis major muscle (cut and reflected)

Ulnar nerve

Coracobrachialis muscle

PLATE **29C** NERVES AND BLOOD VESSELS, THE RIGHT AXILLARY REGION

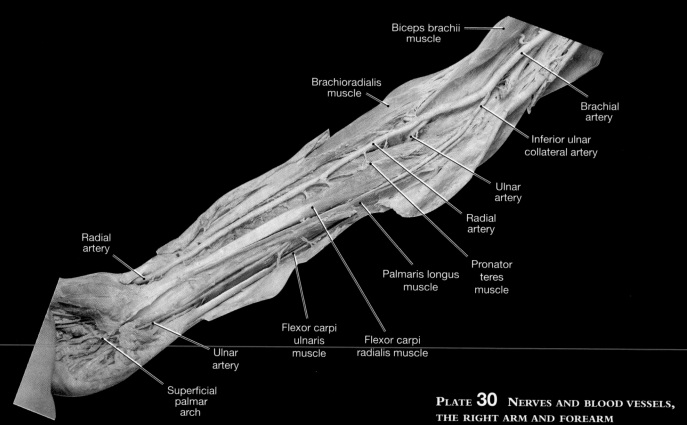

Biceps brachii muscle

Brachioradialis muscle

Brachial artery

Inferior ulnar collateral artery

Ulnar artery

Radial artery

Pronator teres muscle

Palmaris longus muscle

Flexor carpi radialis muscle

Flexor carpi ulnaris muscle

Ulnar artery

Radial artery

Superficial palmar arch

PLATE **30** NERVES AND BLOOD VESSELS, THE RIGHT ARM AND FOREARM

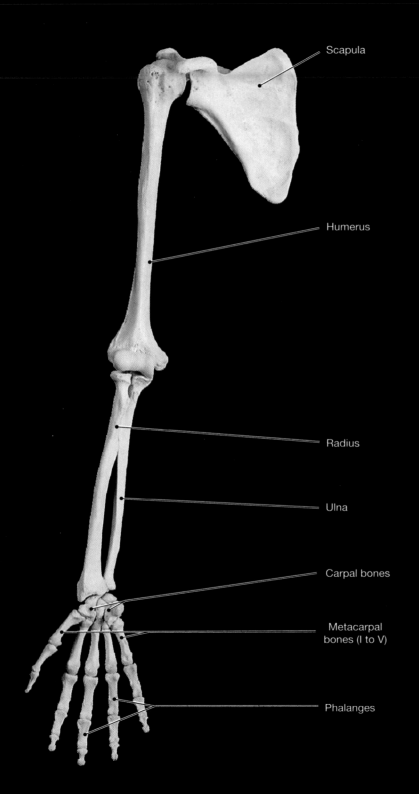

Scapula

Humerus

Radius

Ulna

Carpal bones

Metacarpal
bones (I to V)

Phalanges

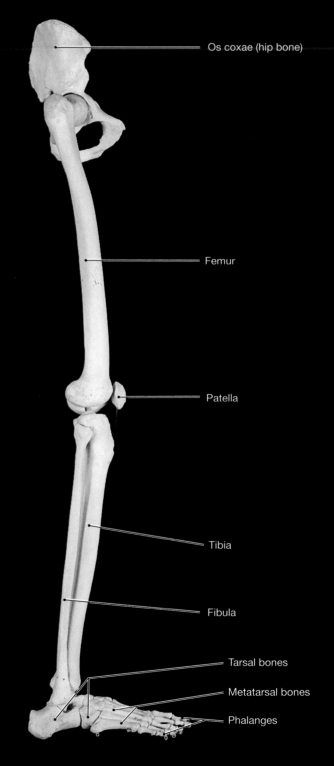

Os coxae (hip bone)

Femur

Patella

Tibia

Fibula

Tarsal bones

Metatarsal bones

Phalanges

PLATE **31** BONES OF THE RIGHT UPPER LIMB, ANTERIOR

PLATE **32** BONES OF THE RIGHT LOWER LIMB, LATERAL

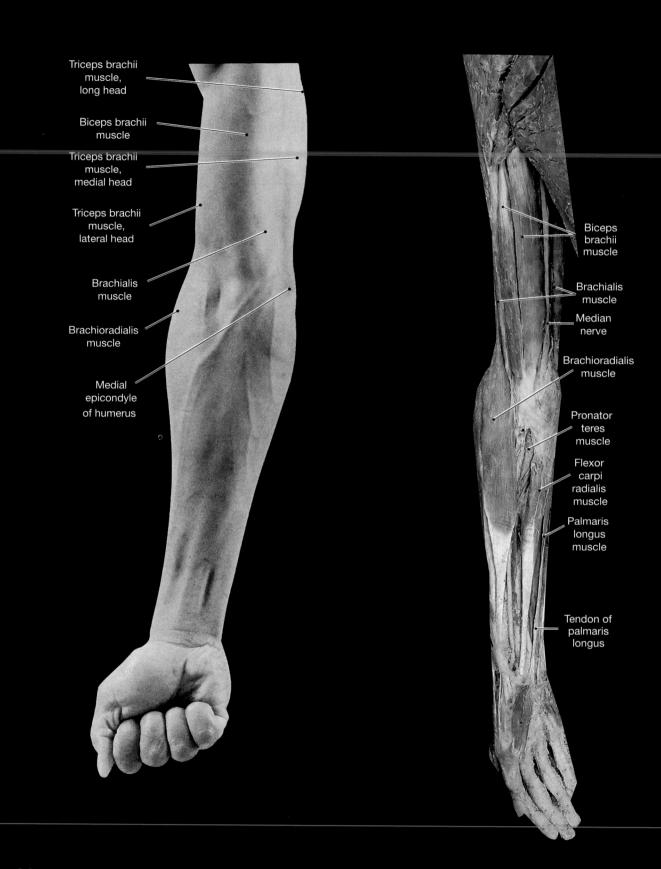

Triceps brachii muscle, long head

Biceps brachii muscle

Triceps brachii muscle, medial head

Triceps brachii muscle, lateral head

Brachialis muscle

Brachioradialis muscle

Medial epicondyle of humerus

Biceps brachii muscle

Brachialis muscle

Median nerve

Brachioradialis muscle

Pronator teres muscle

Flexor carpi radialis muscle

Palmaris longus muscle

Tendon of palmaris longus

PLATE 33a THE RIGHT UPPER LIMB, ANTERIOR SURFACE, MUSCLES

PLATE 33b THE RIGHT UPPER LIMB, ANTERIOR VIEW, SUPERFICIAL DISSECTION

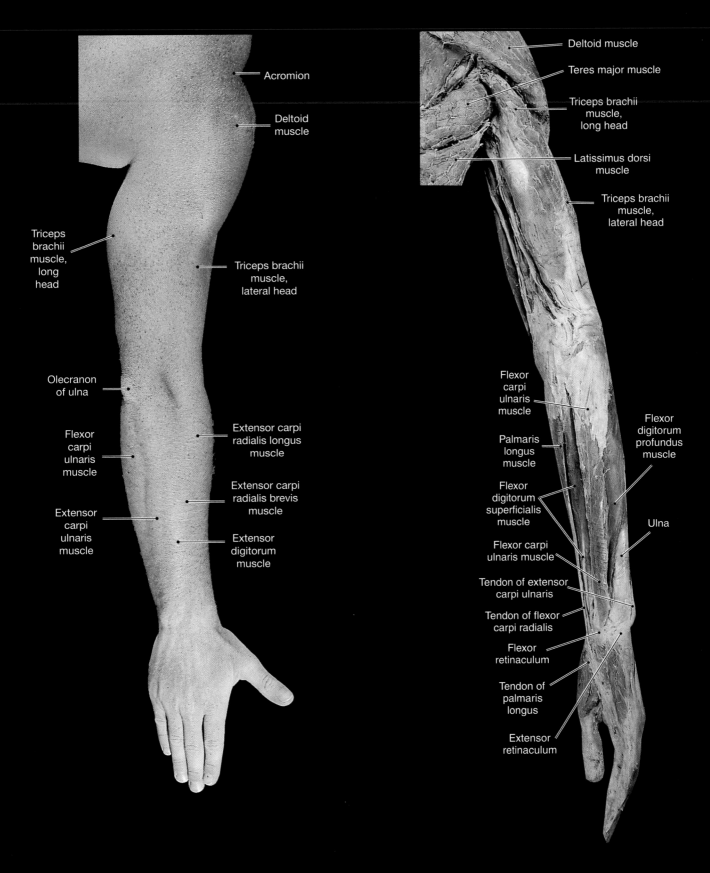

Plate **33c** THE RIGHT UPPER LIMB, POSTERIOR SURFACE, LANDMARKS

Plate **33d** THE RIGHT UPPER LIMB, POSTERIOR VIEW, SUPERFICIAL DISSECTION

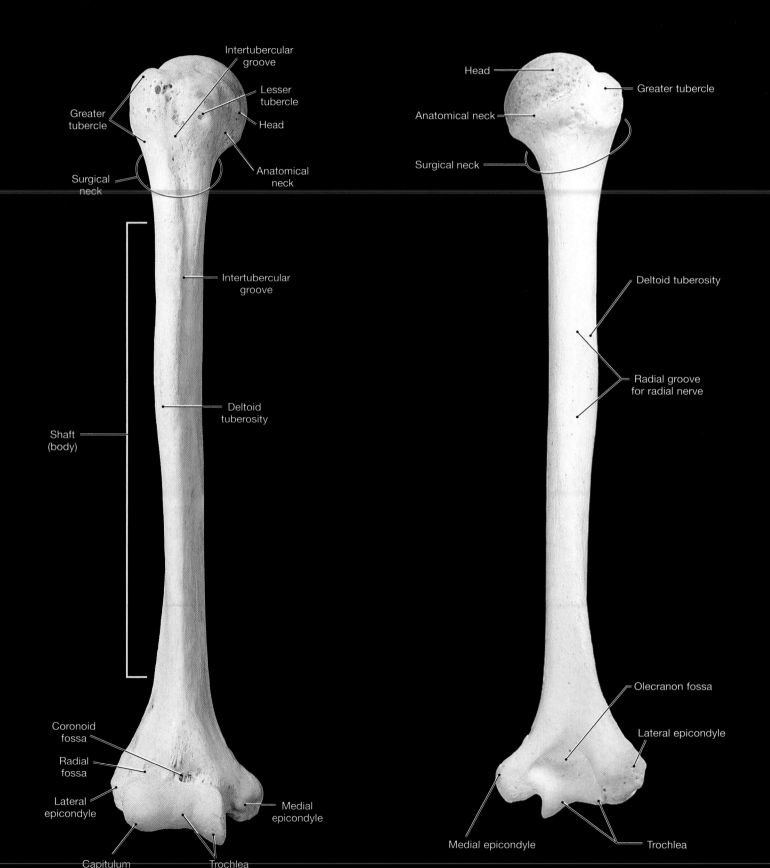

Intertubercular
groove

Lesser
tubercle

Greater
tubercle

Head

Surgical
neck

Anatomical
neck

Intertubercular
groove

Shaft
(body)

Deltoid
tuberosity

Coronoid
fossa

Radial
fossa

Lateral
epicondyle

Medial
epicondyle

Capitulum

Trochlea

Head

Greater tubercle

Anatomical neck

Surgical neck

Deltoid tuberosity

Radial groove
for radial nerve

Olecranon fossa

Lateral epicondyle

Medial epicondyle

Trochlea

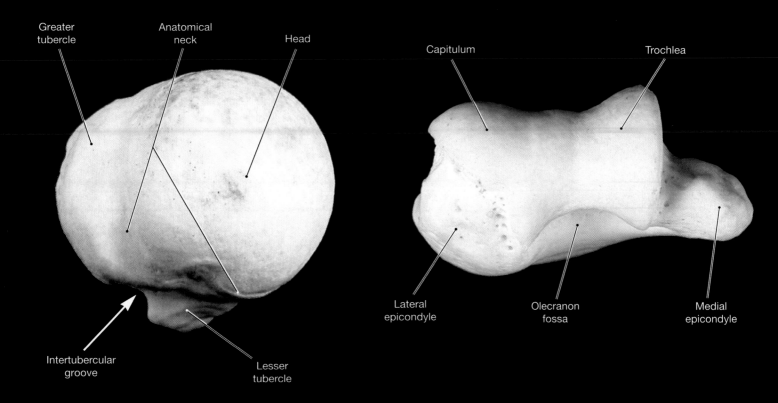

Greater tubercle

Anatomical neck

Head

Intertubercular groove

Lesser tubercle

Capitulum

Trochlea

Lateral epicondyle

Olecranon fossa

Medial epicondyle

PLATE 34c PROXIMAL END OF RIGHT HUMERUS, SUPERIOR VIEW

PLATE 34d DISTAL END OF RIGHT HUMERUS, INFERIOR VIEW

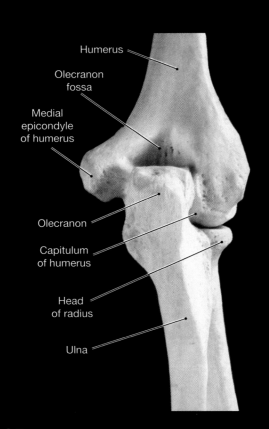

Humerus

Olecranon fossa

Medial epicondyle of humerus

Olecranon

Capitulum of humerus

Head of radius

Ulna

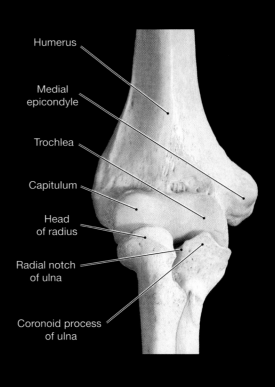

Humerus

Medial epicondyle

Trochlea

Capitulum

Head of radius

Radial notch of ulna

Coronoid process of ulna

PLATE 35a RIGHT ELBOW JOINT, POSTERIOR VIEW

PLATE 35b RIGHT ELBOW JOINT, ANTERIOR VIEW

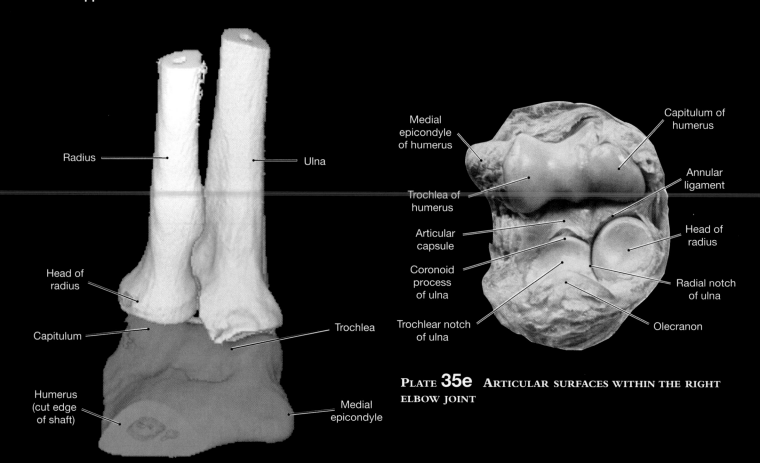

Radius

Ulna

Head of radius

Capitulum

Trochlea

Humerus (cut edge of shaft)

Medial epicondyle

PLATE 35c 3-DIMENSIONAL CT SCAN OF THE RIGHT ELBOW JOINT, SUPERIOR VIEW

Medial epicondyle of humerus

Capitulum of humerus

Trochlea of humerus

Annular ligament

Articular capsule

Head of radius

Coronoid process of ulna

Radial notch of ulna

Trochlear notch of ulna

Olecranon

PLATE 35e ARTICULAR SURFACES WITHIN THE RIGHT ELBOW JOINT

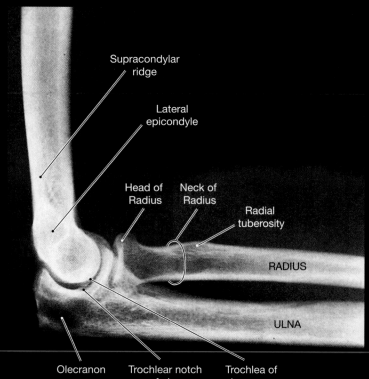

Supracondylar ridge

Lateral epicondyle

Head of Radius

Neck of Radius

Radial tuberosity

RADIUS

ULNA

Olecranon

Trochlear notch of ulna

Trochlea of humerus

PLATE 35d X-RAY OF THE ELBOW JOINT, MEDIAL-LATERAL PROJECTION

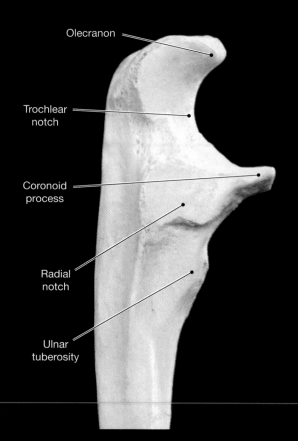

Olecranon

Trochlear notch

Coronoid process

Radial notch

Ulnar tuberosity

PLATE 35f RIGHT ULNA, LATERAL VIEW

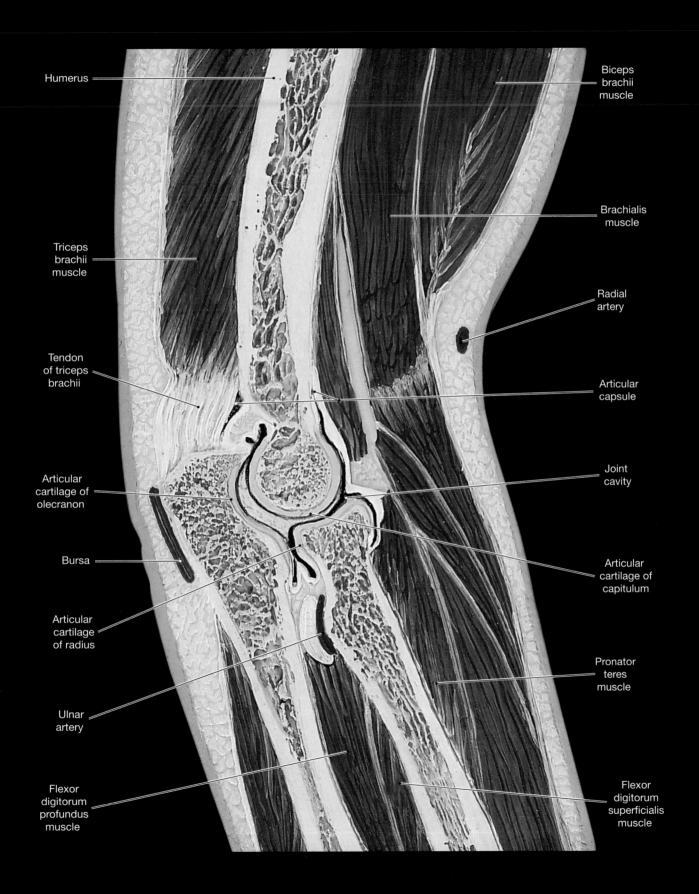

Humerus

Biceps brachii muscle

Brachialis muscle

Triceps brachii muscle

Radial artery

Tendon of triceps brachii

Articular capsule

Articular cartilage of olecranon

Joint cavity

Bursa

Articular cartilage of capitulum

Articular cartilage of radius

Pronator teres muscle

Ulnar artery

Flexor digitorum profundus muscle

Flexor digitorum superficialis muscle

PLATE **35g** THE ELBOW, OBLIQUE SECTION—MODEL

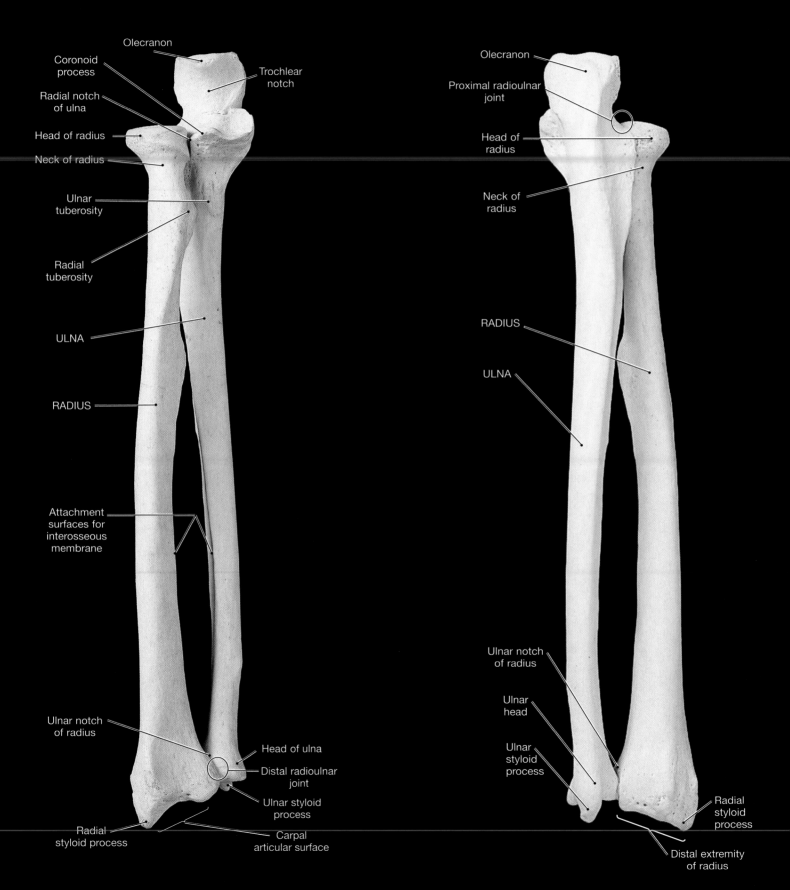

Olecranon

Coronoid process

Trochlear notch

Radial notch of ulna

Head of radius

Neck of radius

Ulnar tuberosity

Radial tuberosity

ULNA

RADIUS

Attachment surfaces for interosseous membrane

Ulnar notch of radius

Head of ulna

Distal radioulnar joint

Ulnar styloid process

Radial styloid process

Carpal articular surface

Olecranon

Proximal radioulnar joint

Head of radius

Neck of radius

RADIUS

ULNA

Ulnar notch of radius

Ulnar head

Ulnar styloid process

Radial styloid process

Distal extremity of radius

PLATE **36a** RIGHT RADIUS AND ULNA, ANTERIOR VIEW

PLATE **36b** RIGHT RADIUS AND ULNA, POSTERIOR VIEW

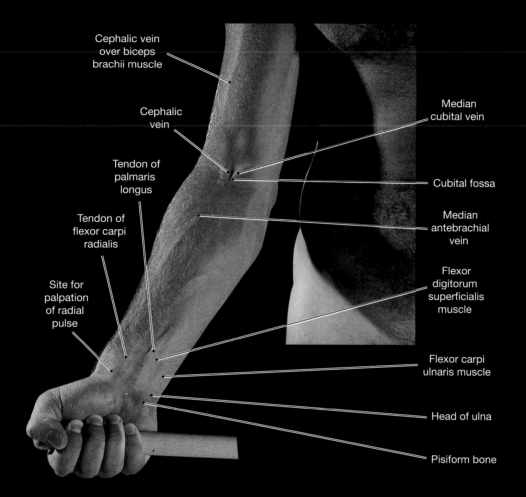

Cephalic vein over biceps brachii muscle

Cephalic vein

Tendon of palmaris longus

Tendon of flexor carpi radialis

Site for palpation of radial pulse

Median cubital vein

Cubital fossa

Median antebrachial vein

Flexor digitorum superficialis muscle

Flexor carpi ulnaris muscle

Head of ulna

Pisiform bone

PLATE 37a THE RIGHT UPPER LIMB, ANTERIOR SURFACE, LANDMARKS

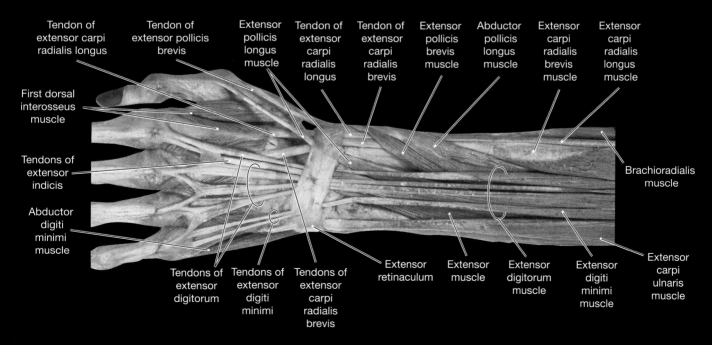

Tendon of extensor carpi radialis longus

Tendon of extensor pollicis brevis

Extensor pollicis longus muscle

Tendon of extensor carpi radialis longus

Tendon of extensor carpi radialis brevis

Extensor pollicis brevis muscle

Abductor pollicis longus muscle

Extensor carpi radialis brevis muscle

Extensor carpi radialis longus muscle

First dorsal interosseus muscle

Tendons of extensor indicis

Abductor digiti minimi muscle

Brachioradialis muscle

Tendons of extensor digitorum

Tendons of extensor digiti minimi

Tendons of extensor carpi radialis brevis

Extensor retinaculum

Extensor muscle

Extensor digitorum muscle

Extensor digiti minimi muscle

Extensor carpi ulnaris muscle

PLATE 37b SUPERFICIAL DISSECTION OF RIGHT FOREARM AND HAND, POSTERIOR VIEW

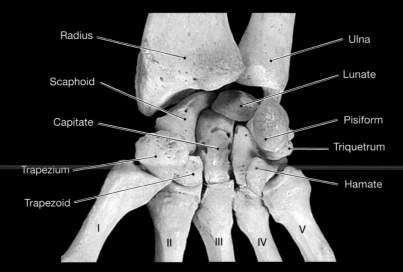

PLATE 38a BONES OF THE RIGHT WRIST, ANTERIOR VIEW

Radius

Ulna

Scaphoid

Lunate

Capitate

Pisiform

Triquetrum

Trapezium

Hamate

Trapezoid

I   II   III   IV   V

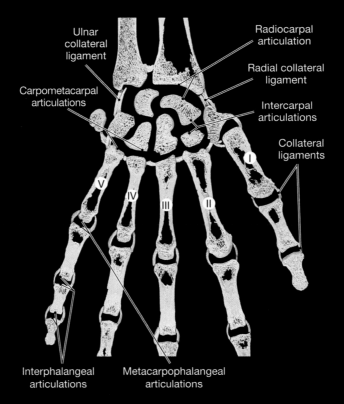

PLATE 38b JOINTS OF THE RIGHT WRIST, CORONAL SECTION

Ulnar collateral ligament

Radiocarpal articulation

Carpometacarpal articulations

Radial collateral ligament

Intercarpal articulations

Collateral ligaments

Interphalangeal articulations

Metacarpophalangeal articulations

PLATE **38c** THE RIGHT HAND,
POSTERIOR VIEW—MODEL

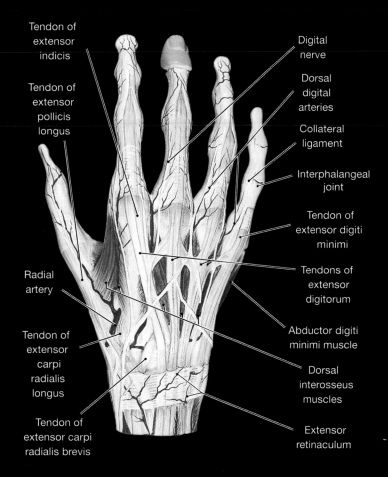

Tendon of extensor indicis

Tendon of extensor pollicis longus

Radial artery

Tendon of extensor carpi radialis longus

Tendon of extensor carpi radialis brevis

Digital nerve

Dorsal digital arteries

Collateral ligament

Interphalangeal joint

Tendon of extensor digiti minimi

Tendons of extensor digitorum

Abductor digiti minimi muscle

Dorsal interosseus muscles

Extensor retinaculum

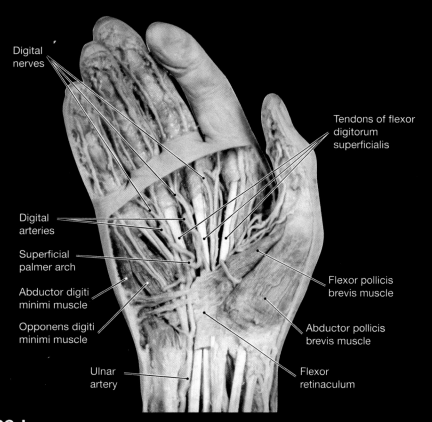

Digital nerves

Digital arteries

Superficial palmer arch

Abductor digiti minimi muscle

Opponens digiti minimi muscle

Ulnar artery

Tendons of flexor digitorum superficialis

Flexor pollicis brevis muscle

Abductor pollicis brevis muscle

Flexor retinaculum

PLATE **38d** SUPERFICIAL DISSECTION OF THE RIGHT WRIST AND HAND, ANTERIOR VIEW

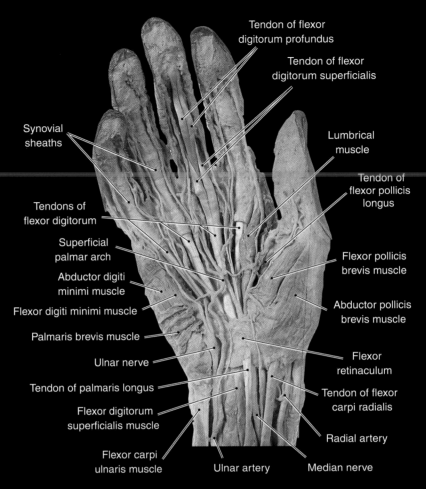

Tendon of flexor
digitorum profundus

Tendon of flexor
digitorum superficialis

Synovial
sheaths

Lumbrical
muscle

Tendon of
flexor pollicis
longus

Tendons of
flexor digitorum

Superficial
palmar arch

Flexor pollicis
brevis muscle

Abductor digiti
minimi muscle

Flexor digiti minimi muscle

Abductor pollicis
brevis muscle

Palmaris brevis muscle

Flexor
retinaculum

Ulnar nerve

Tendon of palmaris longus

Tendon of flexor
carpi radialis

Flexor digitorum
superficialis muscle

Radial artery

Flexor carpi
ulnaris muscle

Ulnar artery

Median nerve

PLATE **38e** THE RIGHT HAND, ANTERIOR VIEW, SUPERFICIAL DISSECTION

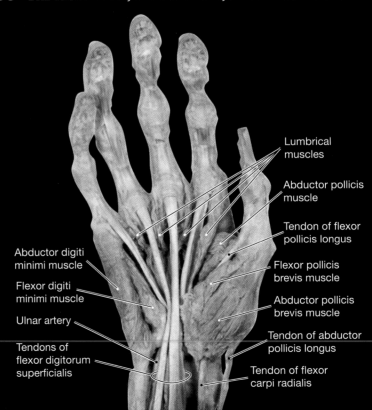

Lumbrical
muscles

Abductor pollicis
muscle

Tendon of flexor
pollicis longus

Abductor digiti
minimi muscle

Flexor pollicis
brevis muscle

Flexor digiti
minimi muscle

Abductor pollicis
brevis muscle

Ulnar artery

Tendon of abductor
pollicis longus

Tendons of
flexor digitorum
superficialis

Tendon of flexor
carpi radialis

PLATE **38f** FLEXOR TENDONS OF RIGHT WRIST AND HAND

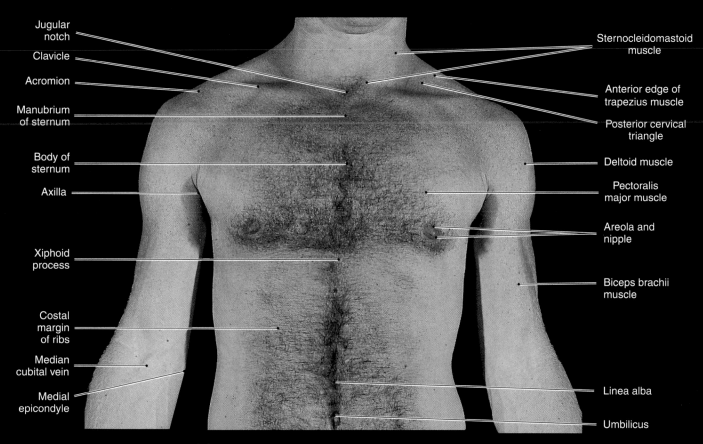

Jugular notch

Clavicle

Acromion

Manubrium of sternum

Body of sternum

Axilla

Xiphoid process

Costal margin of ribs

Median cubital vein

Medial epicondyle

Sternocleidomastoid muscle

Anterior edge of trapezius muscle

Posterior cervical triangle

Deltoid muscle

Pectoralis major muscle

Areola and nipple

Biceps brachii muscle

Linea alba

Umbilicus

PLATE 39a SURFACE ANATOMY OF THE TRUNK, ANTERIOR VIEW

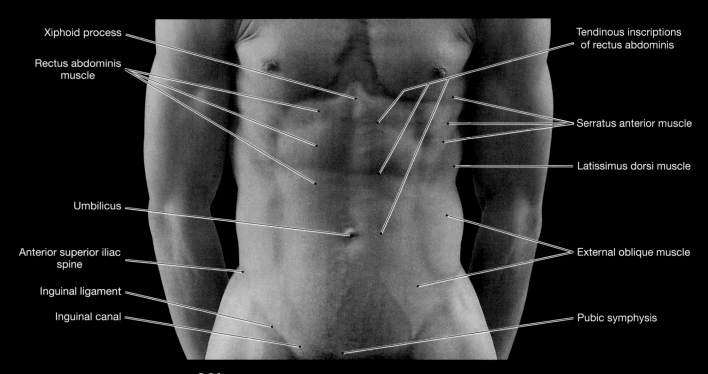

Xiphoid process

Rectus abdominis muscle

Umbilicus

Anterior superior iliac spine

Inguinal ligament

Inguinal canal

Tendinous inscriptions of rectus abdominis

Serratus anterior muscle

Latissimus dorsi muscle

External oblique muscle

Pubic symphysis

PLATE 39b SURFACE ANATOMY OF THE ABDOMEN, ANTERIOR VIEW

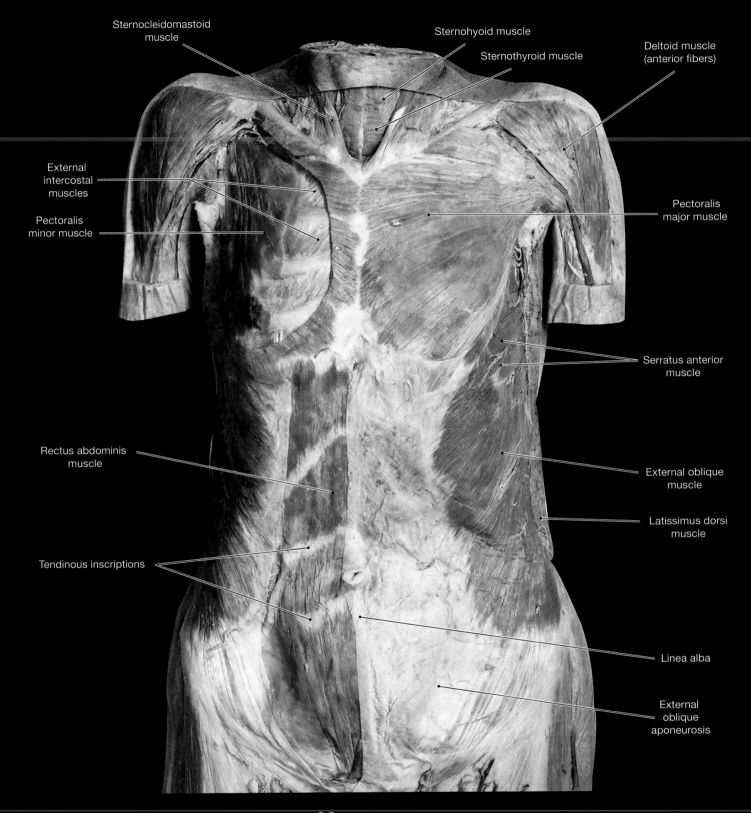

Sternocleidomastoid muscle

Sternohyoid muscle

Sternothyroid muscle

Deltoid muscle (anterior fibers)

External intercostal muscles

Pectoralis minor muscle

Pectoralis major muscle

Serratus anterior muscle

Rectus abdominis muscle

External oblique muscle

Latissimus dorsi muscle

Tendinous inscriptions

Linea alba

External oblique aponeurosis

PLATE 39c   TRUNK, ANTERIOR VIEW

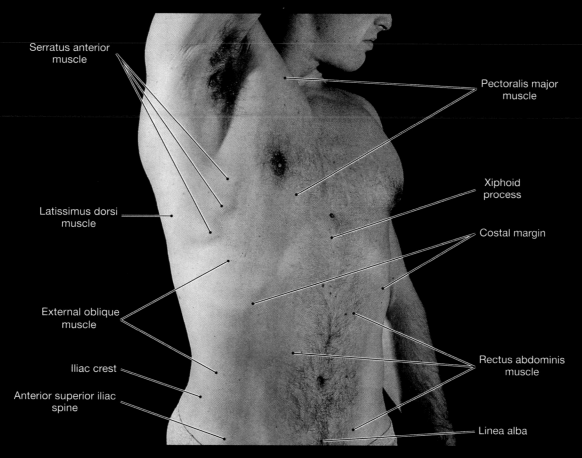

Serratus anterior
muscle

Pectoralis major
muscle

Xiphoid
process

Costal margin

Latissimus dorsi
muscle

External oblique
muscle

Rectus abdominis
muscle

Iliac crest

Anterior superior iliac
spine

Linea alba

PLATE 39d SURFACE ANATOMY OF THE ABDOMEN, ANTEROLATERAL VIEW

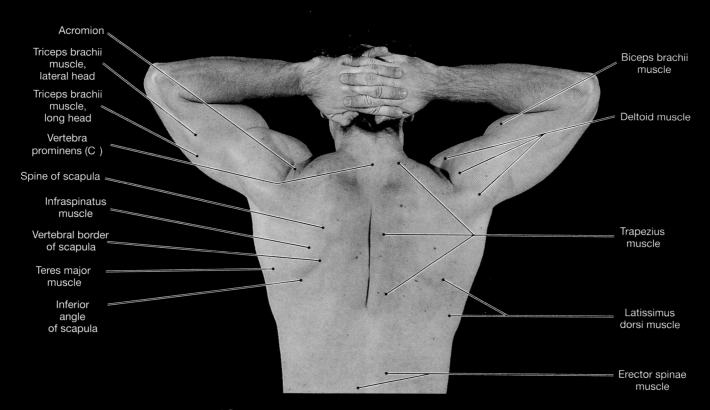

Acromion

Triceps brachii
muscle,
lateral head

Triceps brachii
muscle,
long head

Vertebra
prominens (C)

Spine of scapula

Infraspinatus
muscle

Vertebral border
of scapula

Teres major
muscle

Inferior
angle
of scapula

Biceps brachii
muscle

Deltoid muscle

Trapezius
muscle

Latissimus
dorsi muscle

Erector spinae
muscle

PLATE 40a SURFACE ANATOMY OF THE TRUNK, POSTERIOR VIEW

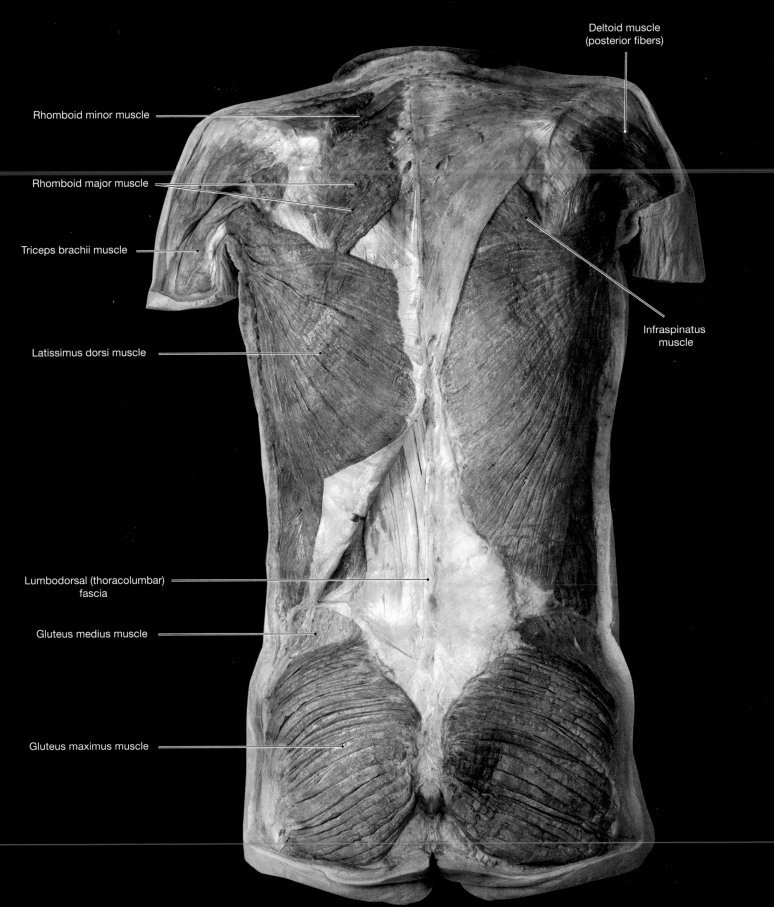

Deltoid muscle
(posterior fibers)

Rhomboid minor muscle

Rhomboid major muscle

Triceps brachii muscle

Infraspinatus
muscle

Latissimus dorsi muscle

Lumbodorsal (thoracolumbar)
fascia

Gluteus medius muscle

Gluteus maximus muscle

PLATE **40b**   TRUNK, POSTERIOR VIEW

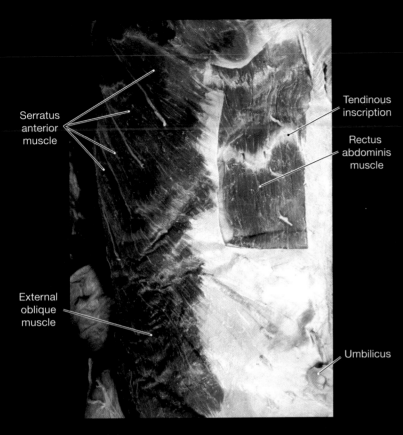

Serratus
anterior
muscle

Tendinous
inscription

Rectus
abdominis
muscle

External
oblique
muscle

Umbilicus

PLATE **41a** ABDOMINAL WALL, ANTERIOR VIEW

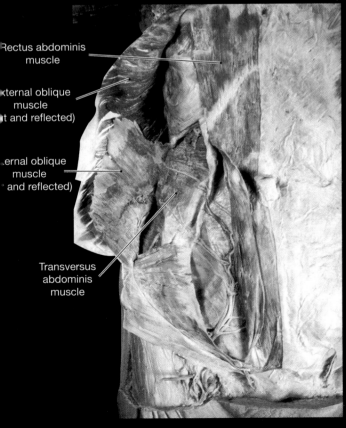

Rectus abdominis
muscle

xternal oblique
muscle
t and reflected)

ernal oblique
muscle
and reflected)

Transversus
abdominis
muscle

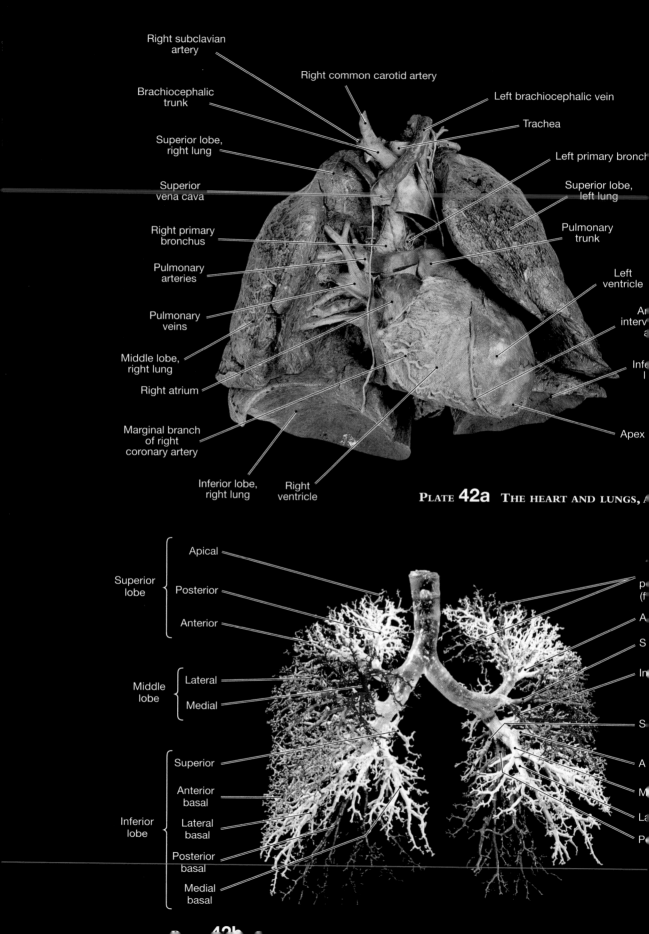

Right subclavian artery

Right common carotid artery

Left brachiocephalic vein

Brachiocephalic trunk

Trachea

Superior lobe, right lung

Left primary bronch

Superior vena cava

Superior lobe, left lung

Right primary bronchus

Pulmonary trunk

Pulmonary arteries

Left ventricle

Pulmonary veins

Ar
interv
a

Middle lobe, right lung

Infe
l

Right atrium

Marginal branch of right coronary artery

Apex

Inferior lobe, right lung

Right ventricle

PLATE **42a** THE HEART AND LUNGS,

Superior lobe

Apical

Posterior

Anterior

p
(f

A

S

Middle lobe

Lateral

Medial

In

S

A

Inferior lobe

Superior

Anterior basal

M

Lateral basal

La

Posterior basal

P

Medial basal

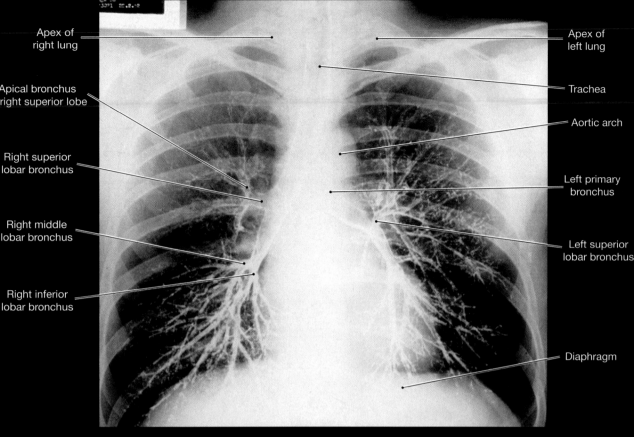

Apex of
right lung

Apical bronchus
right superior lobe

Right superior
lobar bronchus

Right middle
lobar bronchus

Right inferior
lobar bronchus

Apex of
left lung

Trachea

Aortic arch

Left primary
bronchus

Left superior
lobar bronchus

Diaphragm

PLATE **42c** BRONCHOGRAM

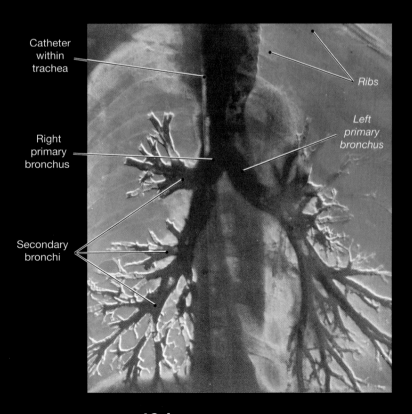

Catheter
within
trachea

Right
primary
bronchus

Secondary
bronchi

Ribs

*Left
primary
bronchus*

PLATE **42d** COLORIZED BRONCHOGRAM

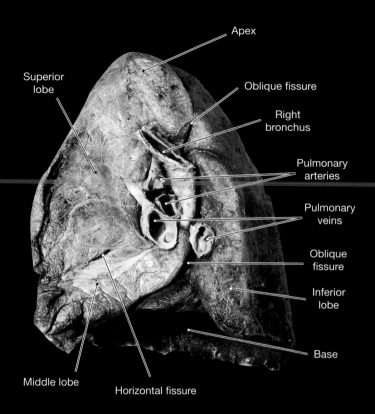

Apex

Superior lobe

Oblique fissure

Right bronchus

Pulmonary arteries

Pulmonary veins

Oblique fissure

Inferior lobe

Base

Middle lobe

Horizontal fissure

PLATE **43a** MEDIAL SURFACE OF RIGHT LUNG

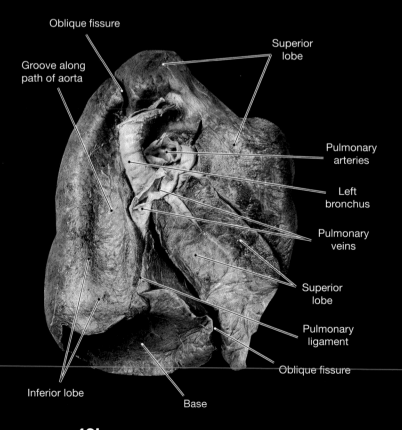

Oblique fissure

Superior lobe

Groove along path of aorta

Pulmonary arteries

Left bronchus

Pulmonary veins

Superior lobe

Pulmonary ligament

Oblique fissure

Inferior lobe

Base

PLATE **43b** MEDIAL SURFACE OF LEFT LUNG

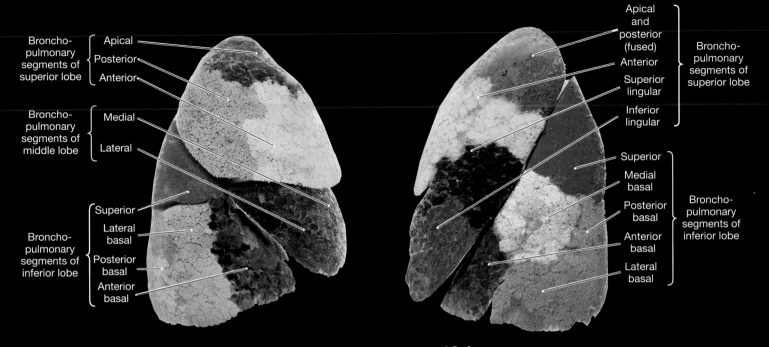

Broncho-pulmonary segments of superior lobe
- Apical
- Posterior
- Anterior

Broncho-pulmonary segments of middle lobe
- Medial
- Lateral

Broncho-pulmonary segments of inferior lobe
- Superior
- Lateral basal
- Posterior basal
- Anterior basal

Broncho-pulmonary segments of superior lobe
- Apical and posterior (fused)
- Anterior
- Superior lingular
- Inferior lingular

Broncho-pulmonary segments of inferior lobe
- Superior
- Medial basal
- Posterior basal
- Anterior basal
- Lateral basal

PLATE **43c** BRONCHOPULMONARY SEGMENTS IN THE RIGHT LUNG, LATERAL VIEW

PLATE **43d** BRONCHOPULMONARY SEGMENTS IN THE LEFT LUNG, LATERAL VIEW

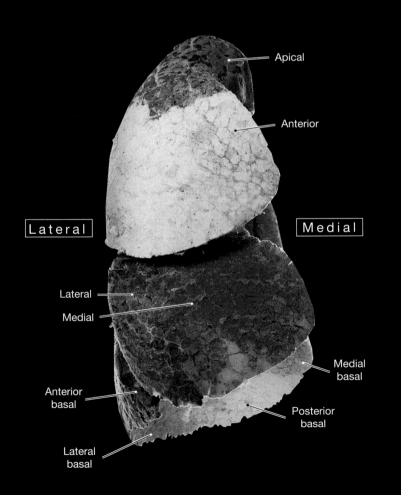

Apical

Anterior

Lateral

Medial

Lateral
Medial

Medial basal

Anterior basal

Posterior basal

Lateral basal

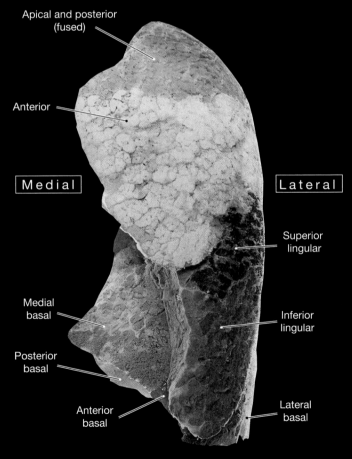

Apical and posterior (fused)

Anterior

Medial

Lateral

Superior lingular

Medial basal

Inferior lingular

Posterior basal

Anterior basal

Lateral basal

PLATE **44a** BRONCHOPULMONARY SEGMENTS IN THE RIGHT LUNG, ANTERIOR VIEW

PLATE **44b** BRONCHOPULMONARY SEGMENTS IN THE LEFT LUNG, ANTERIOR VIEW

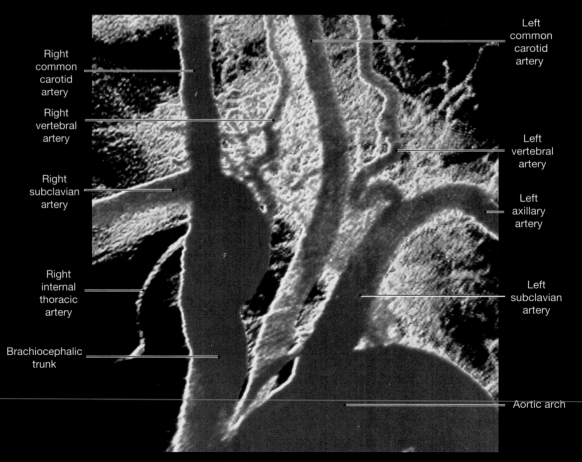

Catheter passing through the right atrium and ventricle to enter the pulmonary trunk

Left pulmonary artery

Right pulmonary artery

Pulmonary trunk

PLATE **44c** PULMONARY ANGIOGRAM

Right common carotid artery

Right vertebral artery

Right subclavian artery

Right internal thoracic artery

Brachiocephalic trunk

Left common carotid artery

Left vertebral artery

Left axillary artery

Left subclavian artery

Aortic arch

PLATE **45a** AORTIC ANGIOGRAM

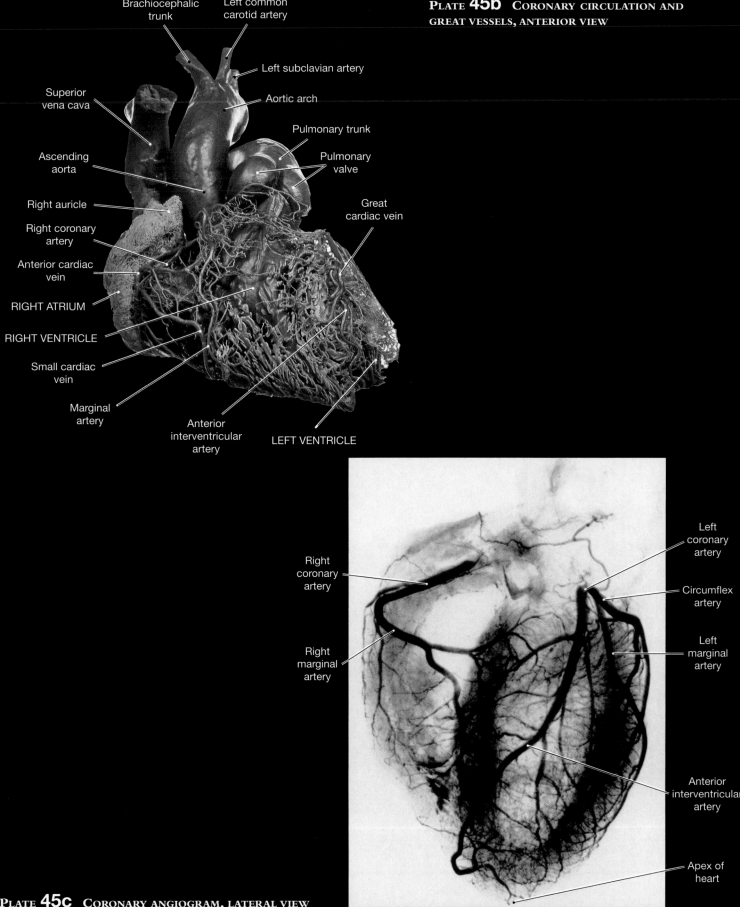

Brachiocephalic trunk

Left common carotid artery

Left subclavian artery

Superior vena cava

Aortic arch

Pulmonary trunk

Pulmonary valve

Ascending aorta

Right auricle

Great cardiac vein

Right coronary artery

Anterior cardiac vein

RIGHT ATRIUM

RIGHT VENTRICLE

Small cardiac vein

Marginal artery

Anterior interventricular artery

LEFT VENTRICLE

Right coronary artery

Left coronary artery

Circumflex artery

Right marginal artery

Left marginal artery

Anterior interventricular artery

Apex of heart

PLATE 45c CORONARY ANGIOGRAM, LATERAL VIEW

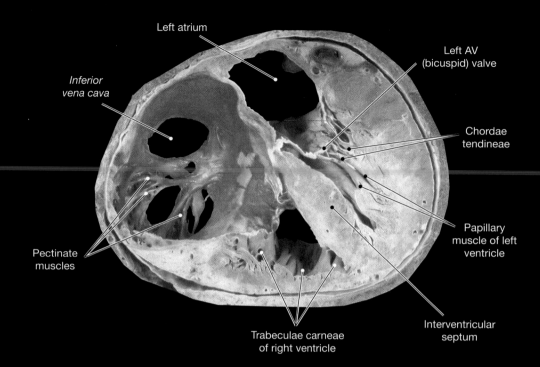

Left atrium

Inferior
vena cava

Left AV
(bicuspid) valve

Chordae
tendineae

Pectinate
muscles

Papillary
muscle of left
ventricle

Trabeculae carneae
of right ventricle

Interventricular
septum

PLATE **45d** HORIZONTAL SECTION THROUGH THE HEART, SUPERIOR VIEW

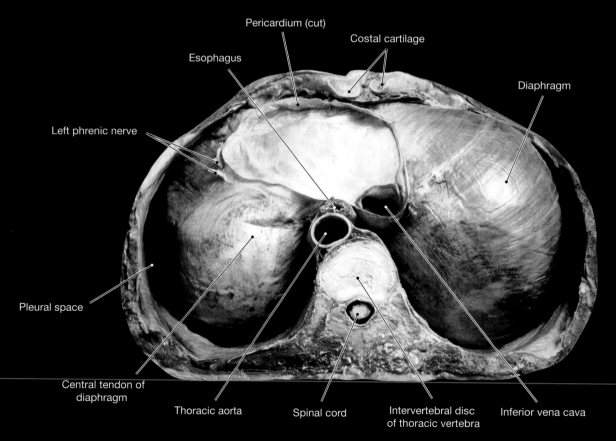

Pericardium (cut)

Esophagus

Costal cartilage

Diaphragm

Left phrenic nerve

Pleural space

Central tendon of
diaphragm

Thoracic aorta

Spinal cord

Intervertebral disc
of thoracic vertebra

Inferior vena cava

PLATE **46** DIAPHRAGM, SUPERIOR VIEW

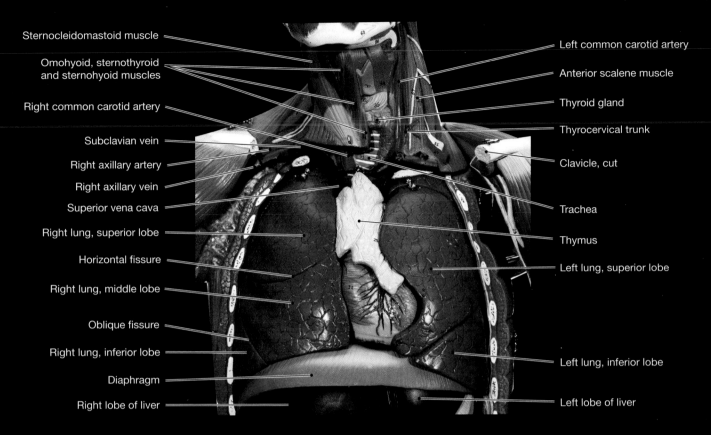

Sternocleidomastoid muscle

Omohyoid, sternothyroid and sternohyoid muscles

Right common carotid artery

Subclavian vein

Right axillary artery

Right axillary vein

Superior vena cava

Right lung, superior lobe

Horizontal fissure

Right lung, middle lobe

Oblique fissure

Right lung, inferior lobe

Diaphragm

Right lobe of liver

Left common carotid artery

Anterior scalene muscle

Thyroid gland

Thyrocervical trunk

Clavicle, cut

Trachea

Thymus

Left lung, superior lobe

Left lung, inferior lobe

Left lobe of liver

PLATE **47a** THORACIC ORGANS, SUPERFICIAL VIEW—MODEL

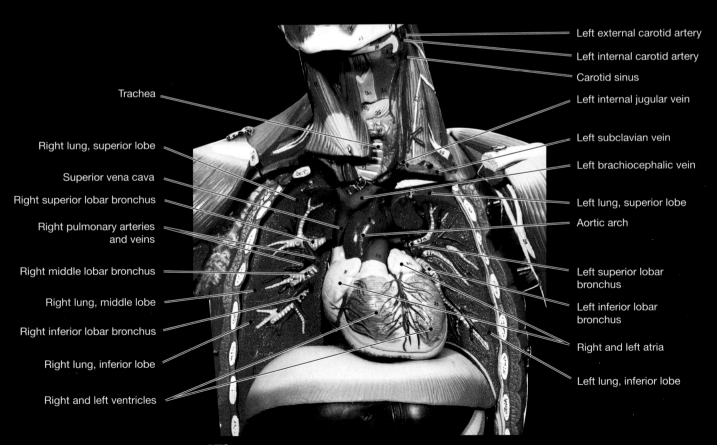

Trachea

Right lung, superior lobe

Superior vena cava

Right superior lobar bronchus

Right pulmonary arteries and veins

Right middle lobar bronchus

Right lung, middle lobe

Right inferior lobar bronchus

Right lung, inferior lobe

Right and left ventricles

Left external carotid artery

Left internal carotid artery

Carotid sinus

Left internal jugular vein

Left subclavian vein

Left brachiocephalic vein

Left lung, superior lobe

Aortic arch

Left superior lobar bronchus

Left inferior lobar bronchus

Right and left atria

Left lung, inferior lobe

PLATE **47b** THORACIC ORGANS, INTERMEDIATE VIEW—MODEL

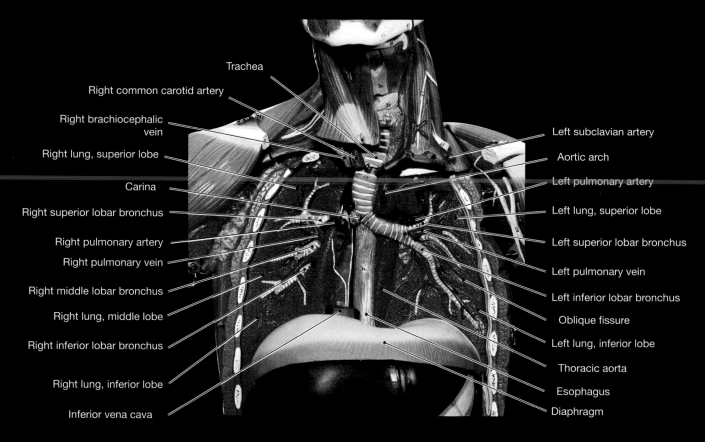

Trachea

Right common carotid artery

Right brachiocephalic vein

Right lung, superior lobe

Carina

Right superior lobar bronchus

Right pulmonary artery

Right pulmonary vein

Right middle lobar bronchus

Right lung, middle lobe

Right inferior lobar bronchus

Right lung, inferior lobe

Inferior vena cava

Left subclavian artery

Aortic arch

Left pulmonary artery

Left lung, superior lobe

Left superior lobar bronchus

Left pulmonary vein

Left inferior lobar bronchus

Oblique fissure

Left lung, inferior lobe

Thoracic aorta

Esophagus

Diaphragm

PLATE **47c** THORACIC ORGANS, DEEPER VIEW—MODEL

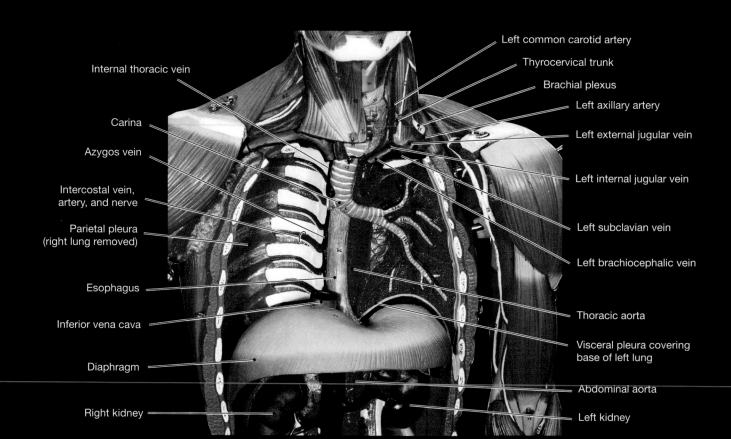

Internal thoracic vein

Carina

Azygos vein

Intercostal vein, artery, and nerve

Parietal pleura (right lung removed)

Esophagus

Inferior vena cava

Diaphragm

Right kidney

Left common carotid artery

Thyrocervical trunk

Brachial plexus

Left axillary artery

Left external jugular vein

Left internal jugular vein

Left subclavian vein

Left brachiocephalic vein

Thoracic aorta

Visceral pleura covering base of left lung

Abdominal aorta

Left kidney

Lymph nodes

Apex of heart

Diaphragm

Descending colon

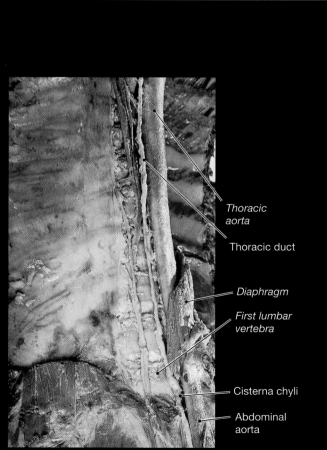

Thoracic aorta

Thoracic duct

Diaphragm

First lumbar vertebra

Cisterna chyli

Abdominal aorta

PLATE **48a** LYMPHATIC TRUNK

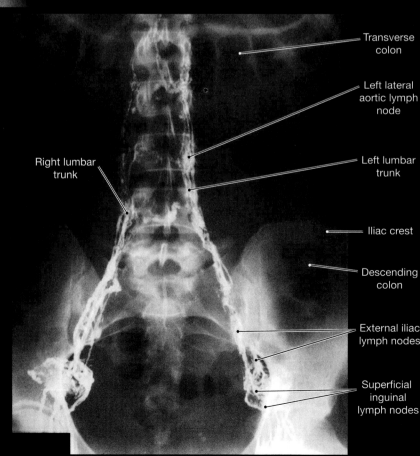

Transverse colon

Left lateral aortic lymph node

Right lumbar trunk

Left lumbar trunk

Iliac crest

Descending colon

External iliac lymph nodes

Superficial inguinal lymph nodes

PLATE **48c** LYMPHANGIOGRAM OF PELVIS, ANTERIOR PROJECTION

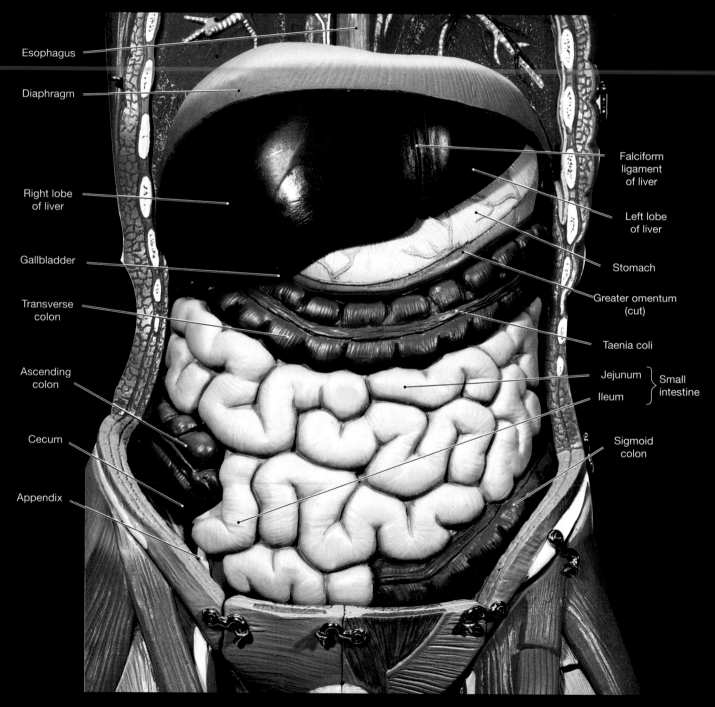

Esophagus

Diaphragm

Right lobe
of liver

Gallbladder

Transverse
colon

Ascending
colon

Cecum

Appendix

Falciform
ligament
of liver

Left lobe
of liver

Stomach

Greater omentum
(cut)

Taenia coli

Jejunum
Ileum } Small
intestine

Sigmoid
colon

PLATE **49a** THE ABDOMINOPELVIC VISCERA, SUPERFICIAL ANTERIOR VIEW

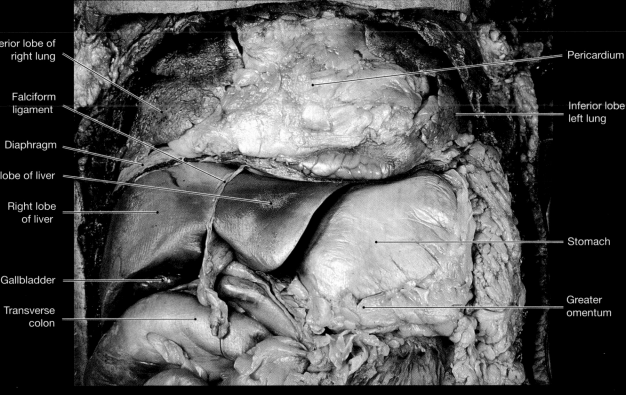

erior lobe of right lung

Falciform ligament

Diaphragm

lobe of liver

Right lobe of liver

Gallbladder

Transverse colon

Pericardium

Inferior lobe left lung

Stomach

Greater omentum

PLATE 49b SUPERIOR PORTION OF ABDOMINOPELVIC CAVITY, ANTERIOR VIEW

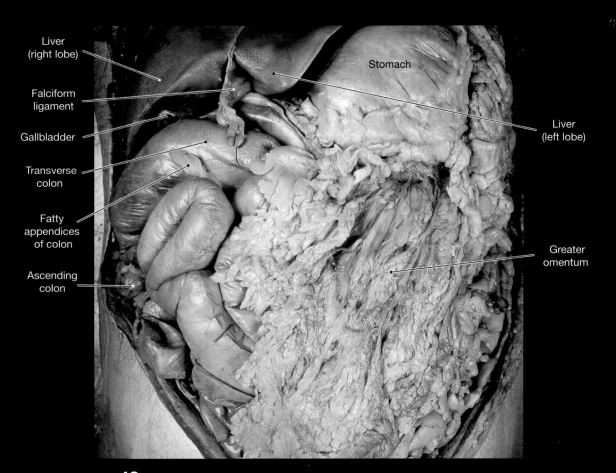

Liver (right lobe)

Falciform ligament

Gallbladder

Transverse colon

Fatty appendices of colon

Ascending colon

Stomach

Liver (left lobe)

Greater omentum

PLATE 49c INFERIOR PORTION OF ABDOMINOPELVIC CAVITY, ANTERIOR VIEW

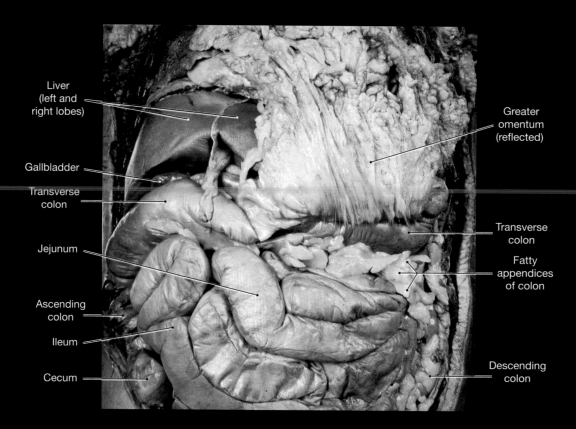

Liver
(left and
right lobes)

Gallbladder

Transverse
colon

Jejunum

Ascending
colon

Ileum

Cecum

Greater
omentum
(reflected)

Transverse
colon

Fatty
appendices
of colon

Descending
colon

PLATE **49d** Abdominal dissection, greater omentum reflected

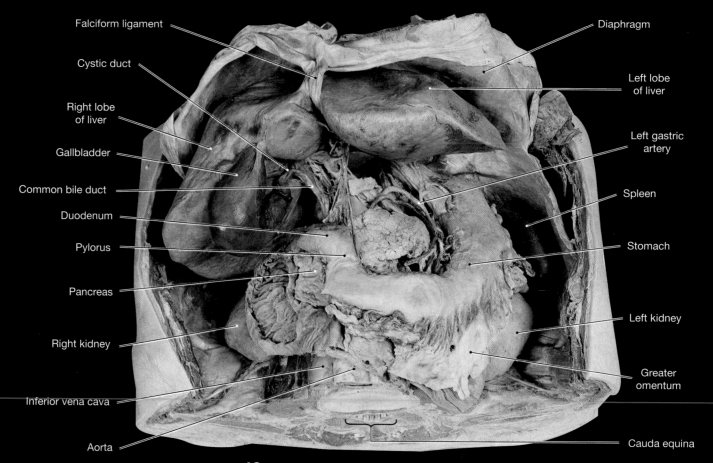

Falciform ligament

Cystic duct

Right lobe
of liver

Gallbladder

Common bile duct

Duodenum

Pylorus

Pancreas

Right kidney

Inferior vena cava

Aorta

Diaphragm

Left lobe
of liver

Left gastric
artery

Spleen

Stomach

Left kidney

Greater
omentum

Cauda equina

PLATE **49e** Liver and gallbladder in situ

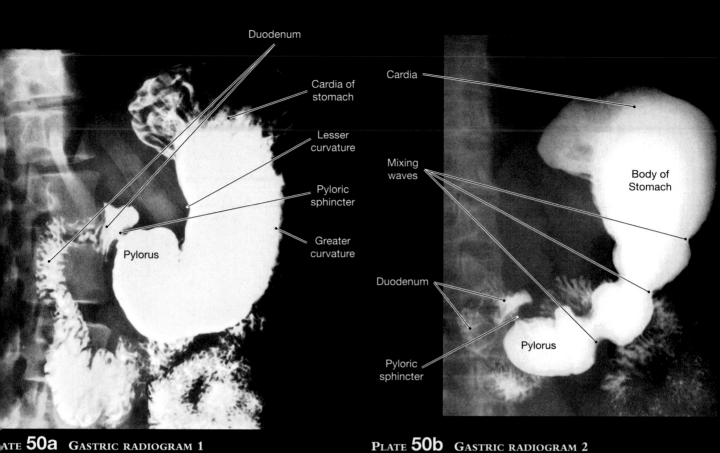

Duodenum

Cardia of stomach

Lesser curvature

Pyloric sphincter

Greater curvature

Pylorus

Cardia

Mixing waves

Body of Stomach

Duodenum

Pylorus

Pyloric sphincter

ᴀᴛᴇ **50a**   Gᴀsᴛʀɪᴄ ʀᴀᴅɪᴏɢʀᴀᴍ 1

Pʟᴀᴛᴇ **50b**   Gᴀsᴛʀɪᴄ ʀᴀᴅɪᴏɢʀᴀᴍ 2

Pylorus

Mixing wave, time 1

Body of stomach

Pylorus

Body of stomach

Mixing waves, time 2

Pʟᴀᴛᴇ **50c**   Gᴀsᴛʀɪᴄ ᴍᴏᴛɪʟɪᴛʏ, ᴀɴᴛᴇʀɪᴏʀ ᴠɪᴇᴡ, ᴛɪᴍᴇs 1 ᴀɴᴅ 2

Right lobe
of liver

Stomach

PLATE **51a** ABDOMINAL DISSECTION, DUODENAL REGION

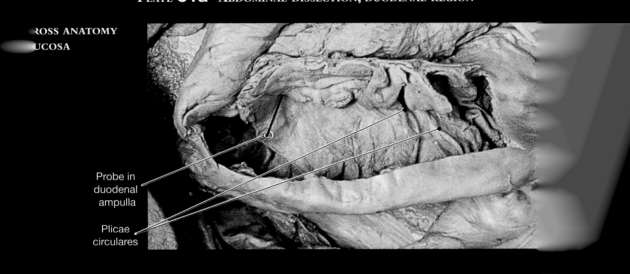

Probe in
duodenal
ampulla

Plicae
circulares

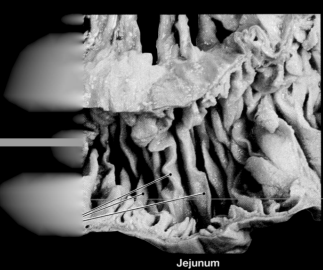

Plicae
circulares

Jejunum

Ileu

..OSS ANATOMY OF THE JEJUNAL MUCOSA

PLATE **51d** CROSS ANATOMY OF TI

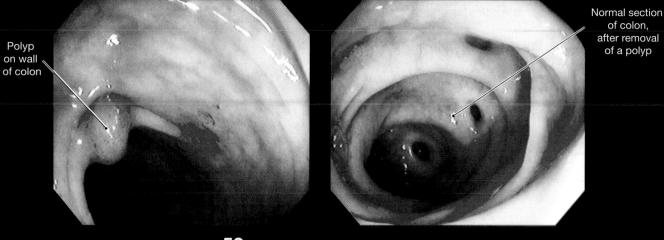

Polyp on wall of colon

Normal section of colon, after removal of a polyp

PLATE 52 NORMAL AND ABNORMAL COLONOSCOPE

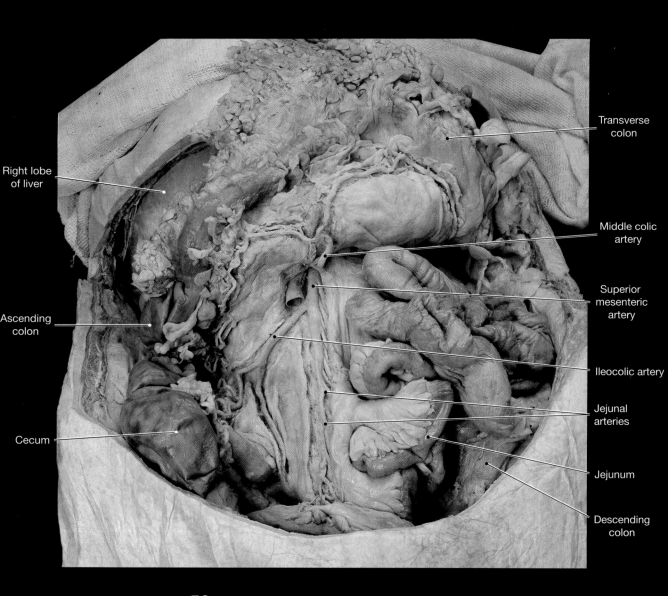

Right lobe of liver

Ascending colon

Cecum

Transverse colon

Middle colic artery

Superior mesenteric artery

Ileocolic artery

Jejunal arteries

Jejunum

Descending colon

PLATE 53a BRANCHES OF THE SUPERIOR MESENTERIC ARTERY

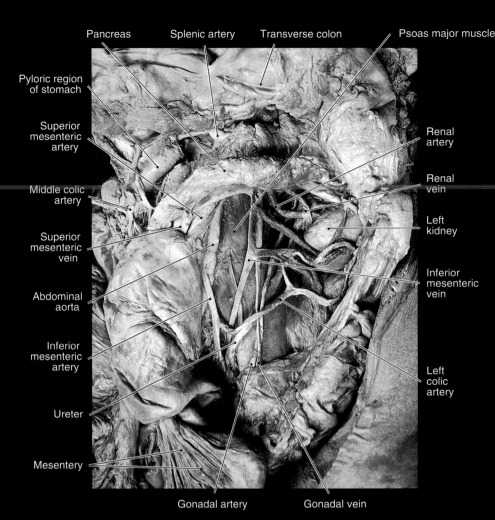

Pancreas   Splenic artery   Transverse colon   Psoas major muscle

Pyloric region
of stomach

Superior
mesenteric
artery

Middle colic
artery

Superior
mesenteric
vein

Abdominal
aorta

Inferior
mesenteric
artery

Ureter

Mesentery

Renal
artery

Renal
vein

Left
kidney

Inferior
mesenteric
vein

Left
colic
artery

Gonadal artery   Gonadal vein

PLATE 53b INFERIOR MESENTERIC VESSELS

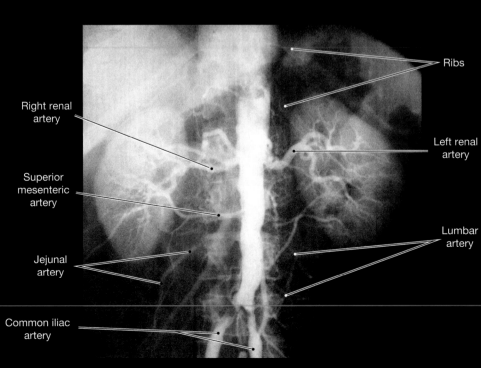

Ribs

Right renal
artery

Left renal
artery

Superior
mesenteric
artery

Lumbar
artery

Jejunal
artery

Common iliac
artery

PLATE 53c ABDOMINAL ARTERIOGRAM 1

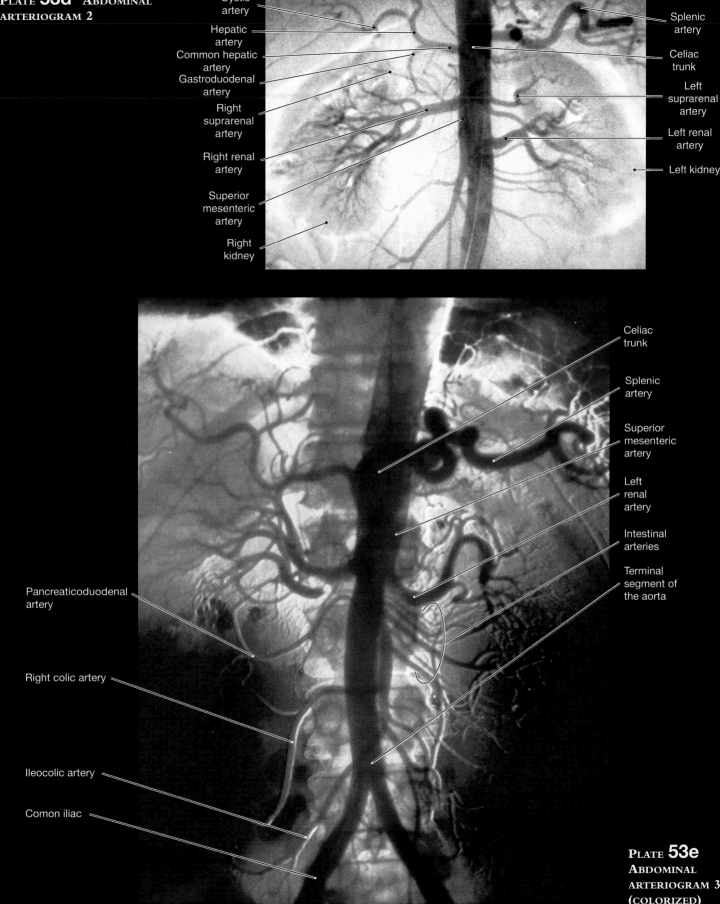

Cystic artery

Hepatic artery

Common hepatic artery

Gastroduodenal artery

Right suprarenal artery

Right renal artery

Superior mesenteric artery

Right kidney

Splenic artery

Celiac trunk

Left suprarenal artery

Left renal artery

Left kidney

Celiac trunk

Splenic artery

Superior mesenteric artery

Left renal artery

Intestinal arteries

Terminal segment of the aorta

Pancreaticoduodenal artery

Right colic artery

Ileocolic artery

Comon iliac

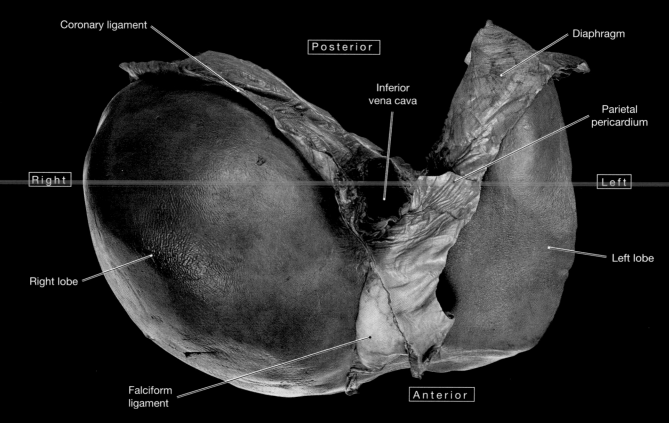

Coronary ligament

Posterior

Inferior
vena cava

Diaphragm

Parietal
pericardium

Right

Left

Right lobe

Left lobe

Falciform
ligament

Anterior

PLATE **54a** THE ISOLATED LIVER AND GALLBLADDER, SUPERIOR VIEW

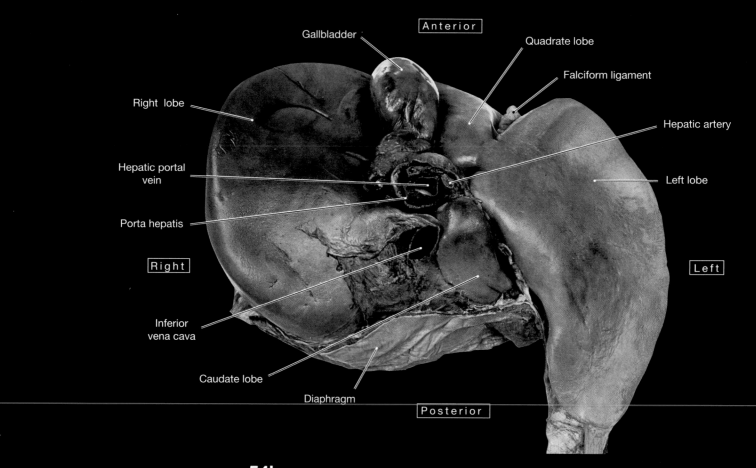

Anterior

Gallbladder

Quadrate lobe

Falciform ligament

Right lobe

Hepatic artery

Hepatic portal
vein

Left lobe

Porta hepatis

Right

Left

Inferior
vena cava

Caudate lobe

Diaphragm

Posterior

PLATE **54b** THE ISOLATED LIVER AND GALLBLADDER, INFERIOR VIEW

PLATE **54c** CORROSION
CAST OF LIVER

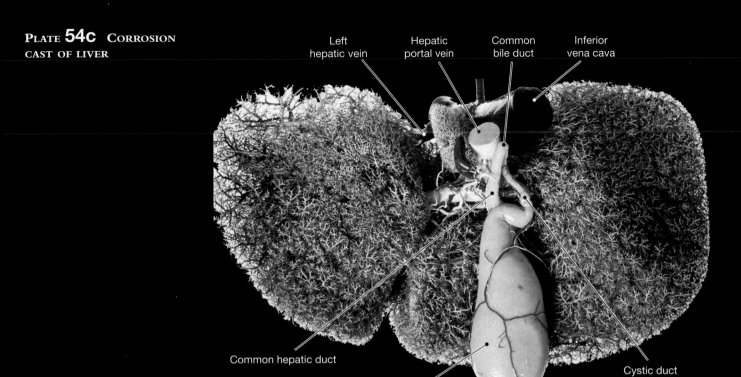

Left hepatic vein

Hepatic portal vein

Common bile duct

Inferior vena cava

Common hepatic duct

Gallbladder

Cystic duct

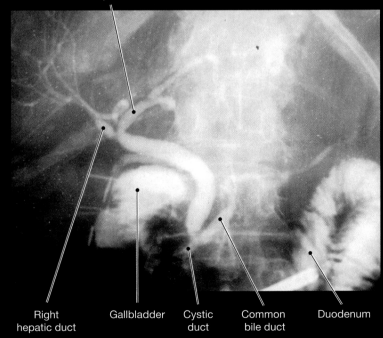

Left hepatic duct

Right hepatic duct

Gallbladder

Cystic duct

Common bile duct

Duodenum

PLATE **54d** CHOLANGIOPAN-CREATOGRAM (PANCREATIC AND BILE DUCTS)

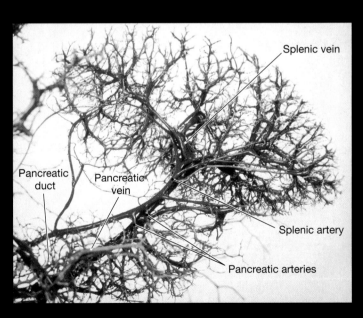

Splenic vein

Pancreatic duct

Pancreatic vein

Splenic artery

Pancreatic arteries

PLATE **55** CORROSION CAST OF SPLENIC AND PANCREATIC VESSELS

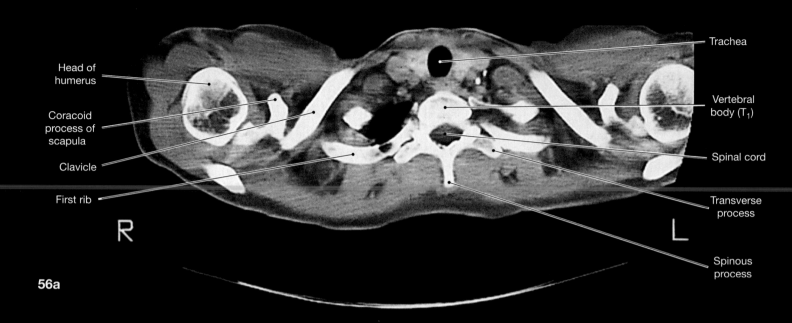

Head of humerus

Coracoid process of scapula

Clavicle

First rib

Trachea

Vertebral body (T₁)

Spinal cord

Transverse process

Spinous process

R

L

56a

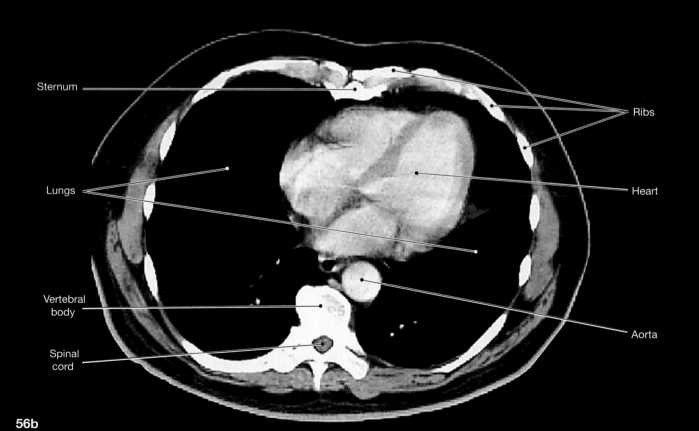

Sternum

Lungs

Vertebral body

Spinal cord

Ribs

Heart

Aorta

56b

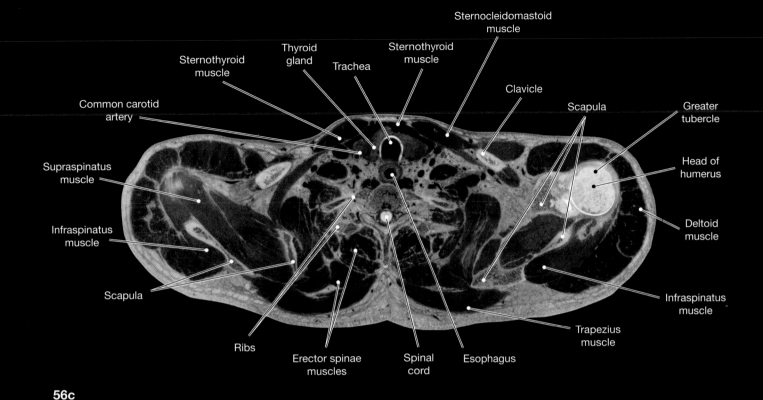

Sternocleidomastoid
muscle

Sternothyroid
muscle

Thyroid
gland

Trachea

Sternothyroid
muscle

Clavicle

Scapula

Greater
tubercle

Common carotid
artery

Supraspinatus
muscle

Head of
humerus

Infraspinatus
muscle

Deltoid
muscle

Scapula

Infraspinatus
muscle

Trapezius
muscle

Ribs

Erector spinae
muscles

Spinal
cord

Esophagus

56c

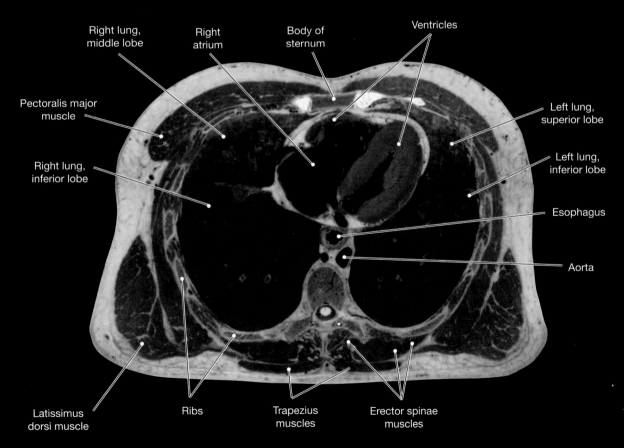

Right lung,
middle lobe

Right
atrium

Body of
sternum

Ventricles

Pectoralis major
muscle

Left lung,
superior lobe

Right lung,
inferior lobe

Left lung,
inferior lobe

Esophagus

Aorta

Latissimus
dorsi muscle

Ribs

Trapezius
muscles

Erector spinae
muscles

56d

PLATES **56c–d** HORIZONTAL SECTIONS THROUGH THE TRUNK

These sections, derived from the Visible Human dataset, are at the approximate levels of the scans shown in parts a–b.

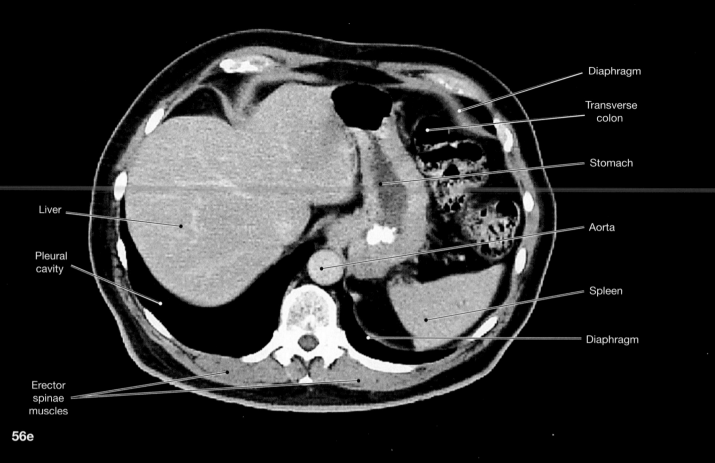

Diaphragm

Transverse colon

Stomach

Liver

Aorta

Pleural cavity

Spleen

Diaphragm

Erector spinae muscles

56e

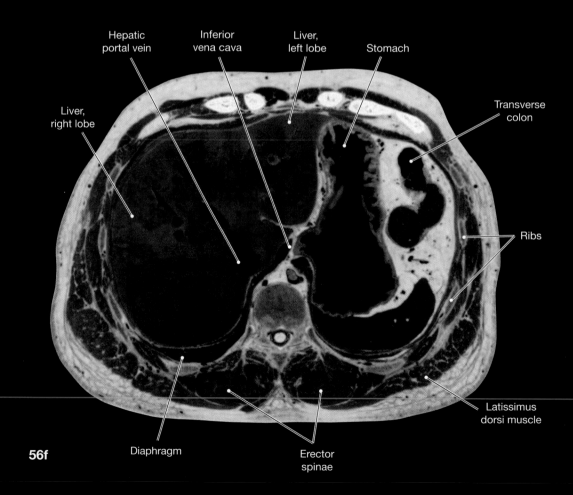

Hepatic portal vein

Inferior vena cava

Liver, left lobe

Stomach

Liver, right lobe

Transverse colon

Ribs

Latissimus dorsi muscle

Diaphragm

Erector spinae

PLATES 56e–f HORIZONTAL
SECTIONS THROUGH THE TRUNK

Section f, derived from the
Visible Human dataset, is at
the approximate levels of the
scan shown in part e.

56f

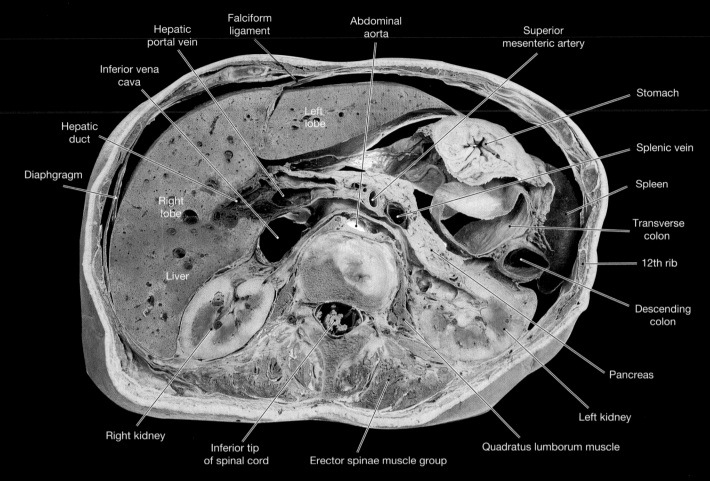

Hepatic portal vein

Falciform ligament

Abdominal aorta

Superior mesenteric artery

Inferior vena cava

Stomach

Hepatic duct

Splenic vein

Diaphgragm

Spleen

Left lobe

Right lobe

Transverse colon

12th rib

Liver

Descending colon

Pancreas

Left kidney

Right kidney

Inferior tip of spinal cord

Erector spinae muscle group

Quadratus lumborum muscle

PLATE 57a ABDOMINAL CAVITY, HORIZONTAL SECTION AT T_{12}

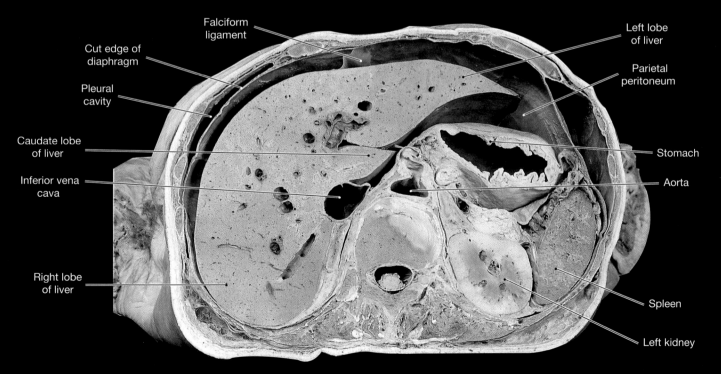

Falciform ligament

Left lobe of liver

Cut edge of diaphragm

Parietal peritoneum

Pleural cavity

Caudate lobe of liver

Stomach

Inferior vena cava

Aorta

Right lobe of liver

Spleen

Left kidney

PLATE 57b ABDOMINAL CAVITY, HORIZONTAL SECTION AT L_1

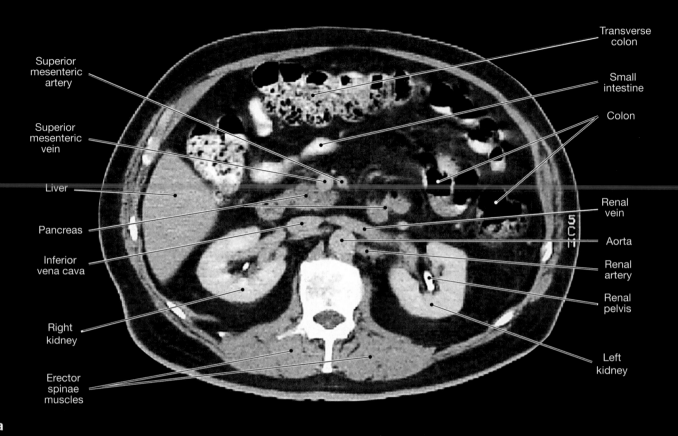

Transverse colon

Superior mesenteric artery

Small intestine

Superior mesenteric vein

Colon

Liver

Renal vein

Pancreas

Aorta

Inferior vena cava

Renal artery

Renal pelvis

Right kidney

Left kidney

Erector spinae muscles

58a

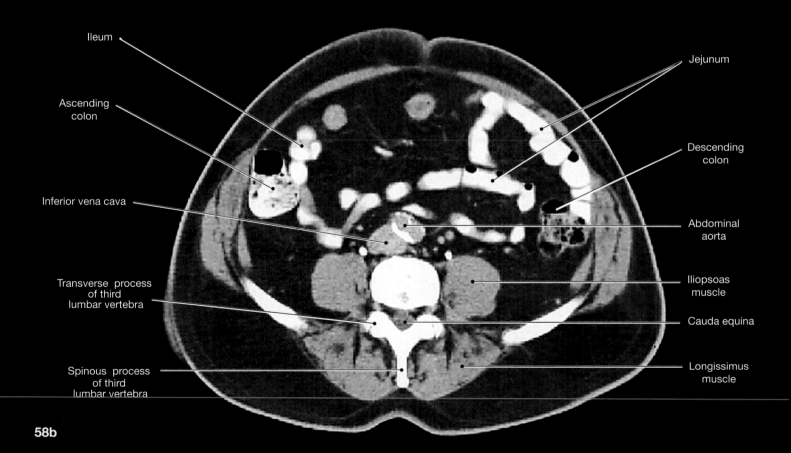

Ileum

Jejunum

Ascending colon

Descending colon

Inferior vena cava

Abdominal aorta

Transverse process of third lumbar vertebra

Iliopsoas muscle

Cauda equina

Spinous process of third lumbar vertebra

Longissimus muscle

58b

PLATES **58a–b** MRI SCANS OF THE TRUNK, HORIZONTAL SECTIONS, SUPERIOR TO INFERIOR SEQUENCE

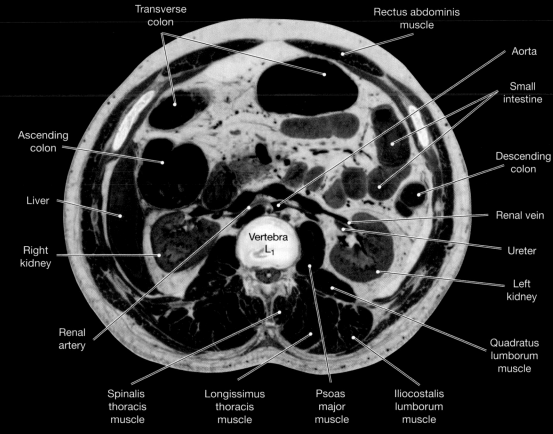

Transverse colon

Rectus abdominis muscle

Aorta

Small intestine

Ascending colon

Descending colon

Liver

Renal vein

Right kidney

Vertebra L₁

Ureter

Left kidney

Renal artery

Quadratus lumborum muscle

Spinalis thoracis muscle

Longissimus thoracis muscle

Psoas major muscle

Iliocostalis lumborum muscle

58c

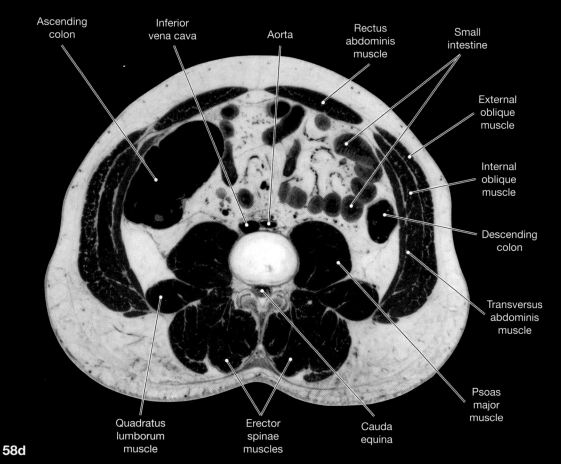

Ascending colon

Inferior vena cava

Aorta

Rectus abdominis muscle

Small intestine

External oblique muscle

Internal oblique muscle

Descending colon

Transversus abdominis muscle

Quadratus lumborum muscle

Erector spinae muscles

Cauda equina

Psoas major muscle

58d

PLATES **58c–d**
HORIZONTAL SECTIONS
THROUGH THE TRUNK

These sections, derived
from the Visible
Human dataset, are at
the approximate levels
of the scans shown in
part a–b.

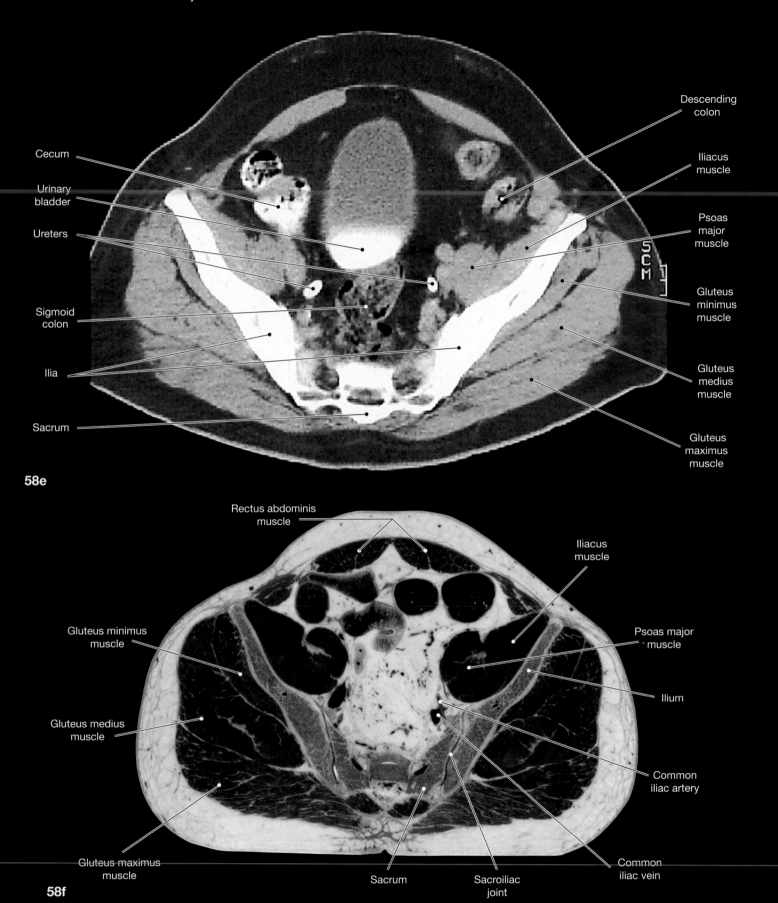

Descending colon

Cecum

Iliacus muscle

Urinary bladder

Psoas major muscle

Ureters

Gluteus minimus muscle

Sigmoid colon

Gluteus medius muscle

Ilia

Sacrum

Gluteus maximus muscle

58e

Rectus abdominis muscle

Iliacus muscle

Gluteus minimus muscle

Psoas major muscle

Ilium

Gluteus medius muscle

Common iliac artery

Gluteus maximus muscle

Sacrum

Sacroiliac joint

Common iliac vein

58f

PLATES 58e–f HORIZONTAL SECTIONS THROUGH THE TRUNK.

Part f, derived from the Visible Human dataset, is at the approximate level of the MRI scan in part e.

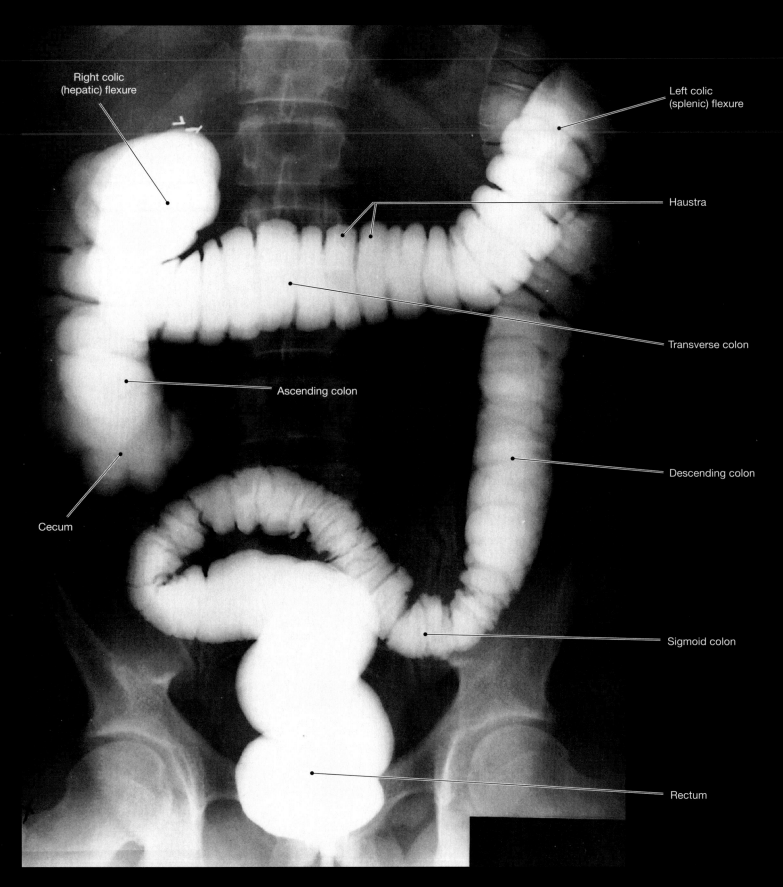

Right colic
(hepatic) flexure

Left colic
(splenic) flexure

Haustra

Transverse colon

Ascending colon

Descending colon

Cecum

Sigmoid colon

Rectum

PLATE **59** CONTRAST X-RAY OF COLON AND RECTUM, ANTERIOR-POSTERIOR PROJECTION

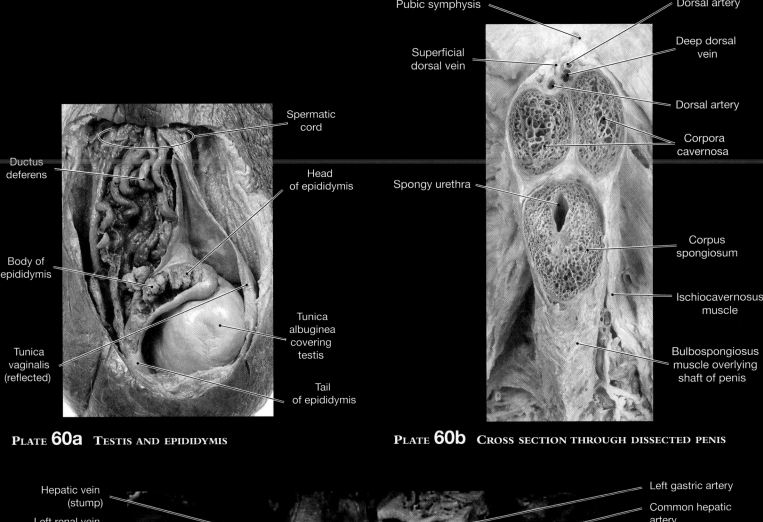

Spermatic cord

Ductus deferens

Head of epididymis

Body of epididymis

Tunica albuginea covering testis

Tunica vaginalis (reflected)

Tail of epididymis

PLATE 60a TESTIS AND EPIDIDYMIS

Pubic symphysis

Dorsal artery

Superficial dorsal vein

Deep dorsal vein

Dorsal artery

Corpora cavernosa

Spongy urethra

Corpus spongiosum

Ischiocavernosus muscle

Bulbospongiosus muscle overlying shaft of penis

PLATE 60b CROSS SECTION THROUGH DISSECTED PENIS

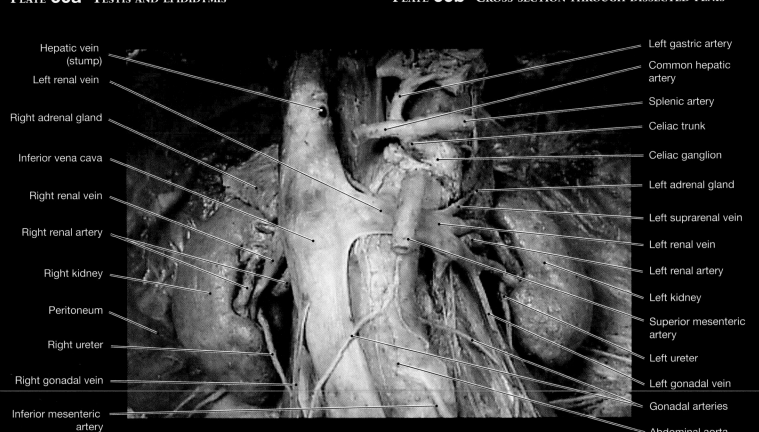

Hepatic vein (stump)

Left renal vein

Right adrenal gland

Inferior vena cava

Right renal vein

Right renal artery

Right kidney

Peritoneum

Right ureter

Right gonadal vein

Inferior mesenteric artery

Left gastric artery

Common hepatic artery

Splenic artery

Celiac trunk

Celiac ganglion

Left adrenal gland

Left suprarenal vein

Left renal vein

Left renal artery

Left kidney

Superior mesenteric artery

Left ureter

Left gonadal vein

Gonadal arteries

Abdominal aorta

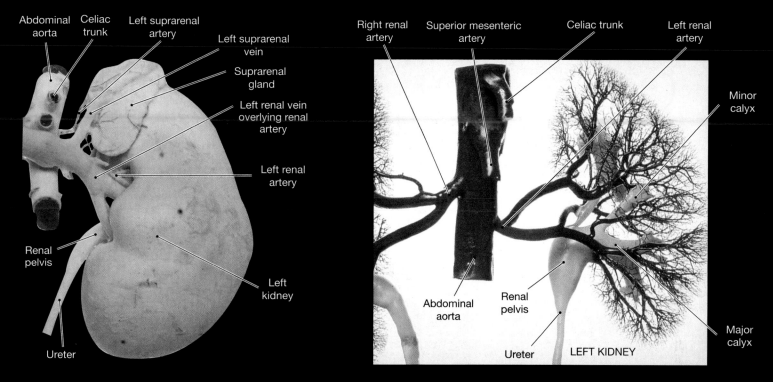

Abdominal aorta

Celiac trunk

Left suprarenal artery

Left suprarenal vein

Suprarenal gland

Left renal vein overlying renal artery

Left renal artery

Renal pelvis

Left kidney

Ureter

PLATE 61b LEFT KIDNEY, ANTERIOR VIEW

Right renal artery

Superior mesenteric artery

Celiac trunk

Left renal artery

Minor calyx

Abdominal aorta

Renal pelvis

Ureter

LEFT KIDNEY

Major calyx

PLATE 61c CORROSION CAST OF THE RENAL ARTERIES, URETERS, AND RENAL PELVIS

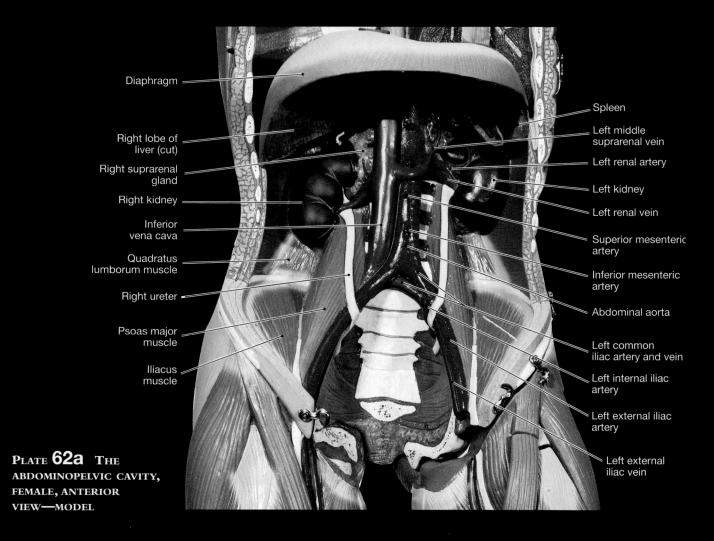

Diaphragm

Right lobe of liver (cut)

Right suprarenal gland

Right kidney

Inferior vena cava

Quadratus lumborum muscle

Right ureter

Psoas major muscle

Iliacus muscle

Spleen

Left middle suprarenal vein

Left renal artery

Left kidney

Left renal vein

Superior mesenteric artery

Inferior mesenteric artery

Abdominal aorta

Left common iliac artery and vein

Left internal iliac artery

Left external iliac artery

Left external iliac vein

PLATE 62a THE ABDOMINOPELVIC CAVITY, FEMALE, ANTERIOR VIEW—MODEL

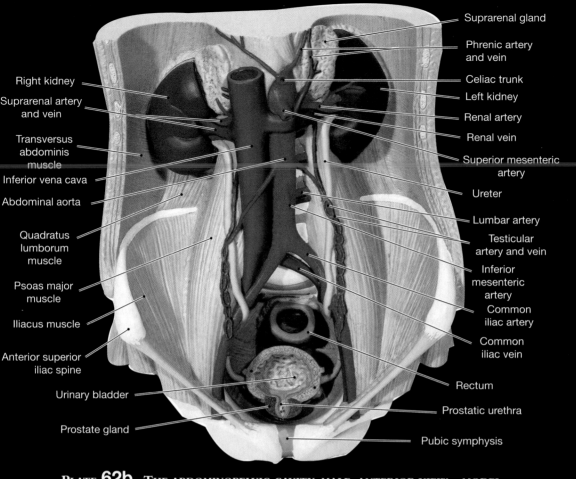

Suprarenal gland

Phrenic artery
and vein

Celiac trunk

Left kidney

Renal artery

Renal vein

Superior mesenteric
artery

Ureter

Lumbar artery

Testicular
artery and vein

Inferior
mesenteric
artery

Common
iliac artery

Common
iliac vein

Rectum

Prostatic urethra

Pubic symphysis

Right kidney

Suprarenal artery
and vein

Transversus
abdominis
muscle

Inferior vena cava

Abdominal aorta

Quadratus
lumborum
muscle

Psoas major
muscle

Iliacus muscle

Anterior superior
iliac spine

Urinary bladder

Prostate gland

PLATE **62b** THE ABDOMINOPELVIC CAVITY, MALE, ANTERIOR VIEW—MODEL

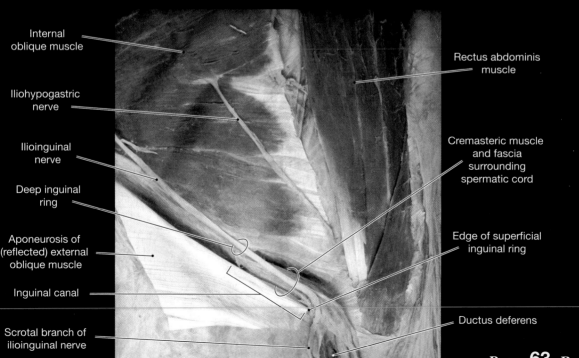

Internal
oblique muscle

Iliohypogastric
nerve

Ilioinguinal
nerve

Deep inguinal
ring

Aponeurosis of
(reflected) external
oblique muscle

Inguinal canal

Scrotal branch of
ilioinguinal nerve

Rectus abdominis
muscle

Cremasteric muscle
and fascia
surrounding
spermatic cord

Edge of superficial
inguinal ring

Ductus deferens

PLATE **63** RIGHT
QUADRANT (RLQ).

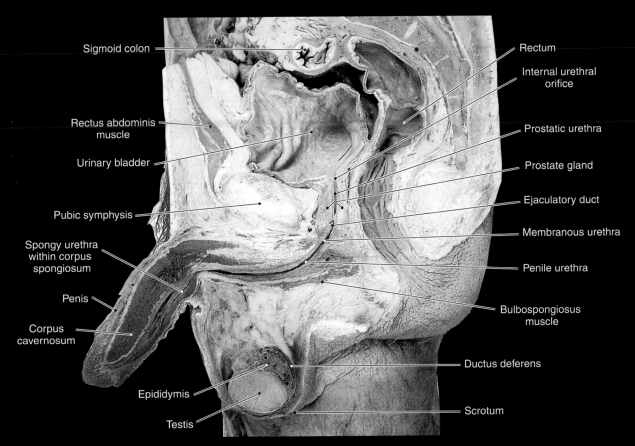

Sigmoid colon

Rectus abdominis muscle

Urinary bladder

Pubic symphysis

Spongy urethra within corpus spongiosum

Penis

Corpus cavernosum

Epididymis

Testis

Rectum

Internal urethral orifice

Prostatic urethra

Prostate gland

Ejaculatory duct

Membranous urethra

Penile urethra

Bulbospongiosus muscle

Ductus deferens

Scrotum

PLATE **64** PELVIC REGION OF A MALE, SAGITTAL SECTION

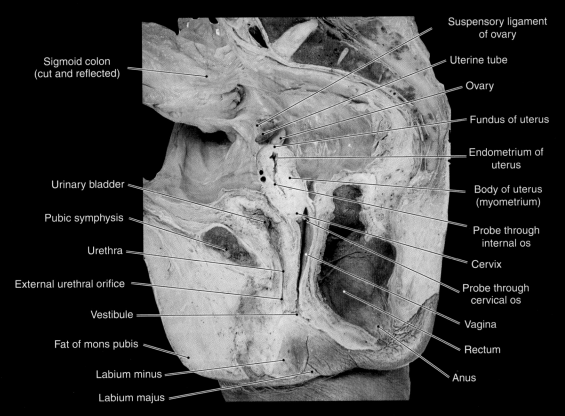

Sigmoid colon (cut and reflected)

Urinary bladder

Pubic symphysis

Urethra

External urethral orifice

Vestibule

Fat of mons pubis

Labium minus

Labium majus

Suspensory ligament of ovary

Uterine tube

Ovary

Fundus of uterus

Endometrium of uterus

Body of uterus (myometrium)

Probe through internal os

Cervix

Probe through cervical os

Vagina

Rectum

Anus

PLATE **65** PELVIC REGION OF A FEMALE, SAGITTAL SECTION

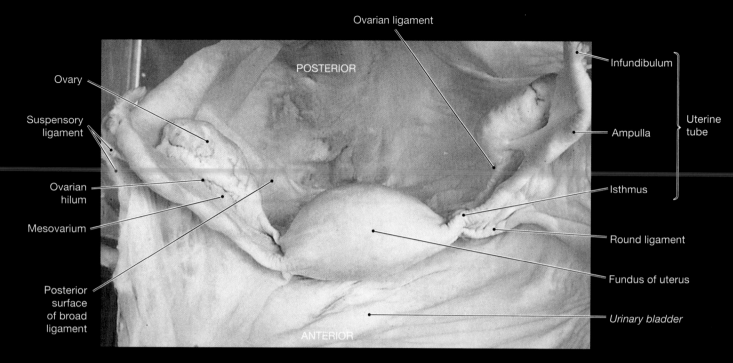

Ovarian ligament

Ovary

Suspensory ligament

POSTERIOR

Infundibulum

Ampulla

Uterine tube

Ovarian hilum

Mesovarium

Isthmus

Round ligament

Fundus of uterus

Posterior surface of broad ligament

Urinary bladder

ANTERIOR

PLATE 66   PELVIC CAVITY, SUPERIOR VIEW

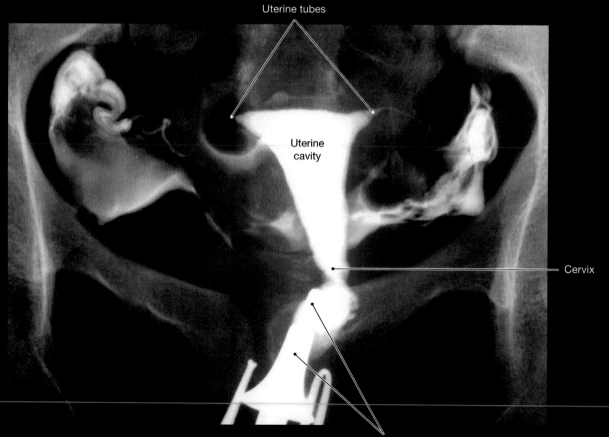

Uterine tubes

Uterine cavity

Cervix

Application tube in vagina, source of contrast medium

PLATE 67   HYSTEROSALPINGOGRAM

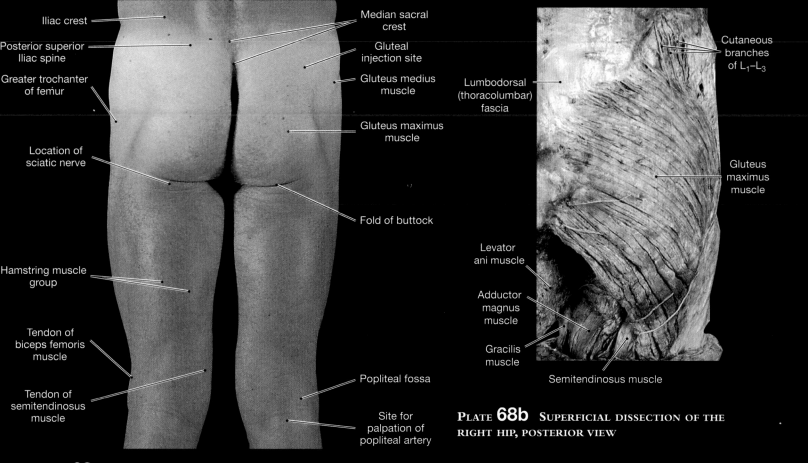

Iliac crest

Posterior superior Iliac spine

Greater trochanter of femur

Location of sciatic nerve

Hamstring muscle group

Tendon of biceps femoris muscle

Tendon of semitendinosus muscle

Median sacral crest

Gluteal injection site

Gluteus medius muscle

Gluteus maximus muscle

Fold of buttock

Popliteal fossa

Site for palpation of popliteal artery

PLATE 68a GLUTEAL REGION AND THIGHS

Lumbodorsal (thoracolumbar) fascia

Cutaneous branches of L_1–L_3

Gluteus maximus muscle

Levator ani muscle

Adductor magnus muscle

Gracilis muscle

Semitendinosus muscle

PLATE 68b SUPERFICIAL DISSECTION OF THE RIGHT HIP, POSTERIOR VIEW

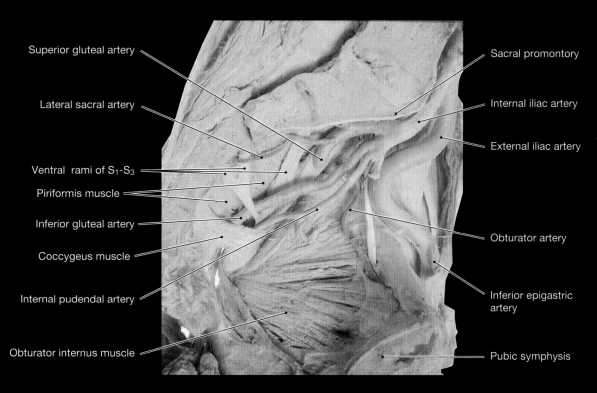

Superior gluteal artery

Lateral sacral artery

Ventral rami of S_1-S_3

Piriformis muscle

Inferior gluteal artery

Coccygeus muscle

Internal pudendal artery

Obturator internus muscle

Sacral promontory

Internal iliac artery

External iliac artery

Obturator artery

Inferior epigastric artery

Pubic symphysis

PLATE 68c BLOOD VESSELS, NERVES, AND MUSCLES IN THE LEFT HALF OF THE PELVIS

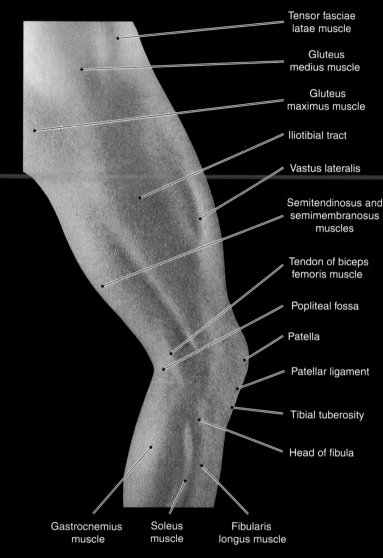

Tensor fasciae
latae muscle

Gluteus
medius muscle

Gluteus
maximus muscle

Iliotibial tract

Vastus lateralis

Semitendinosus and
semimembranosus
muscles

Tendon of biceps
femoris muscle

Popliteal fossa

Patella

Patellar ligament

Tibial tuberosity

Head of fibula

Gastrocnemius   Soleus   Fibularis
muscle   muscle   longus muscle

**PLATE 69a SURFACE ANATOMY OF THE THIGH,
LATERAL VIEW**

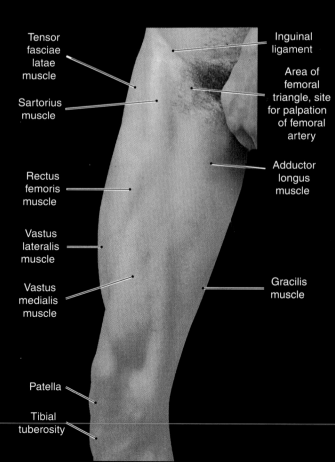

Tensor
fasciae
latae
muscle

Sartorius
muscle

Rectus
femoris
muscle

Vastus
lateralis
muscle

Vastus
medialis
muscle

Patella

Tibial
tuberosity

Inguinal
ligament

Area of
femoral
triangle, site
for palpation
of femoral
artery

Adductor
longus
muscle

Gracilis
muscle

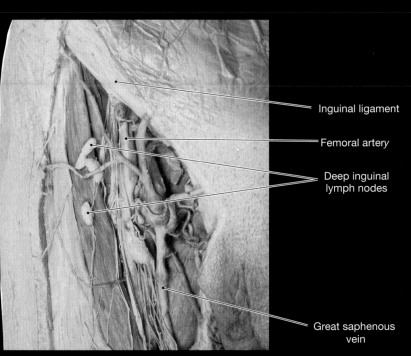

Inguinal ligament

Femoral artery

Deep inguinal lymph nodes

Great saphenous vein

PLATE 70a DISSECTION OF THE RIGHT INGUINAL REGION, MALE

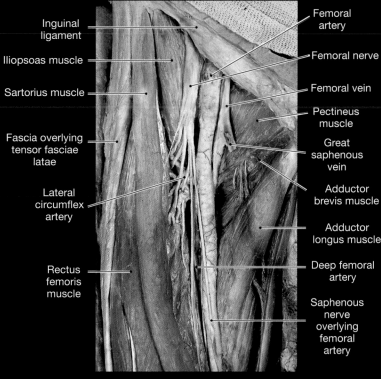

Inguinal ligament

Iliopsoas muscle

Sartorius muscle

Fascia overlying tensor fasciae latae

Lateral circumflex artery

Rectus femoris muscle

Femoral artery

Femoral nerve

Femoral vein

Pectineus muscle

Great saphenous vein

Adductor brevis muscle

Adductor longus muscle

Deep femoral artery

Saphenous nerve overlying femoral artery

PLATE 70b SUPERFICIAL DISSECTION OF RIGHT THIGH, ANTERIOR VIEW

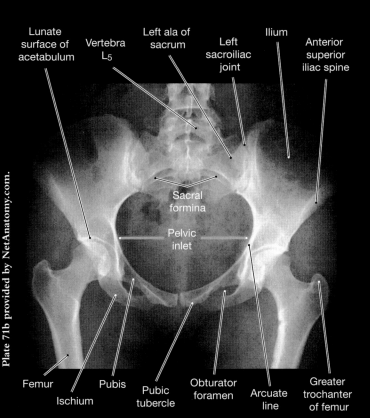

Plate 71b provided by NetAnatomy.com.

Lunate surface of acetabulum

Vertebra L5

Left ala of sacrum

Left sacroiliac joint

Ilium

Anterior superior iliac spine

Sacral formina

Pelvic inlet

Femur

Ischium

Pubis

Pubic tubercle

Obturator foramen

Arcuate line

Greater trochanter of femur

PLATE 71a X-RAY OF PELVIS AND PROXIMAL FEMORA, ANTERIOR-POSTERIOR PROJECTION

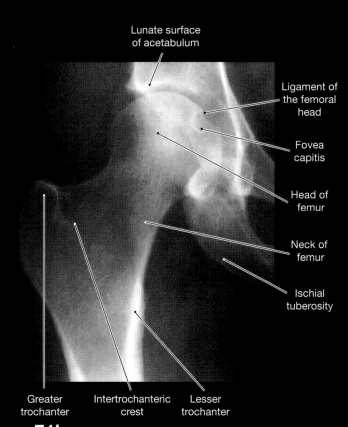

Lunate surface of acetabulum

Ligament of the femoral head

Fovea capitis

Head of femur

Neck of femur

Ischial tuberosity

Greater trochanter

Intertrochanteric crest

Lesser trochanter

PLATE 71b X-RAY OF THE RIGHT HIP JOINT, ANTERIOR-POSTERIOR PROJECTION

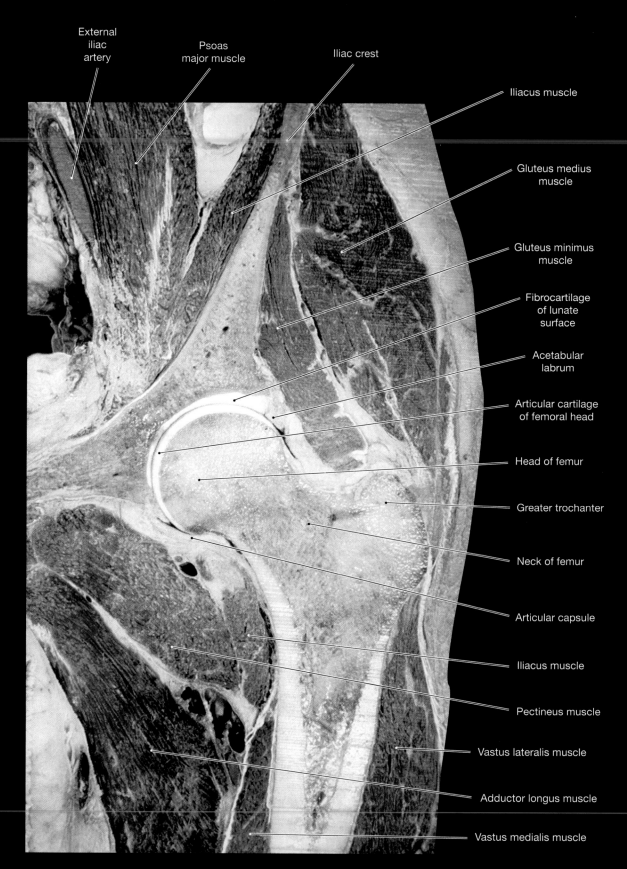

External
iliac
artery

Psoas
major muscle

Iliac crest

Iliacus muscle

Gluteus medius
muscle

Gluteus minimus
muscle

Fibrocartilage
of lunate
surface

Acetabular
labrum

Articular cartilage
of femoral head

Head of femur

Greater trochanter

Neck of femur

Articular capsule

Iliacus muscle

Pectineus muscle

Vastus lateralis muscle

Adductor longus muscle

Vastus medialis muscle

PLATE **72a** CORONAL SECTION THROUGH THE LEFT HIP

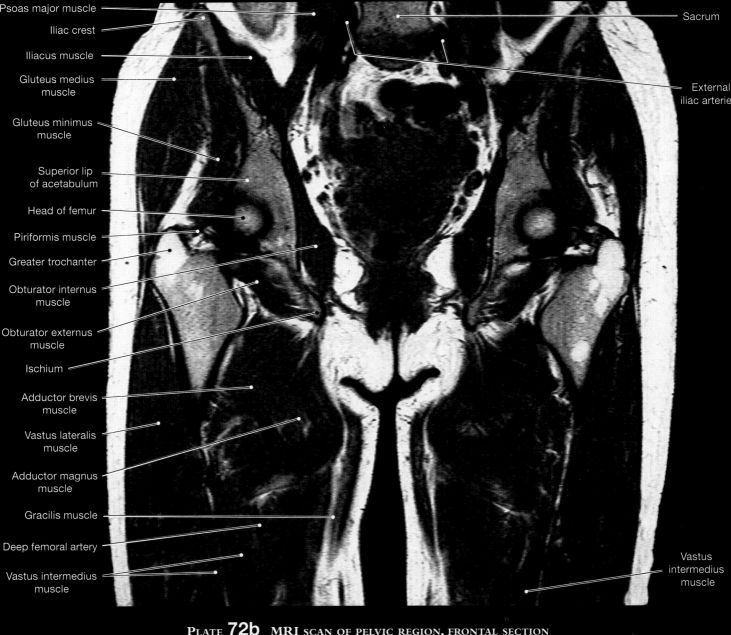

Psoas major muscle

Iliac crest

Iliacus muscle

Gluteus medius muscle

Gluteus minimus muscle

Superior lip of acetabulum

Head of femur

Piriformis muscle

Greater trochanter

Obturator internus muscle

Obturator externus muscle

Ischium

Adductor brevis muscle

Vastus lateralis muscle

Adductor magnus muscle

Gracilis muscle

Deep femoral artery

Vastus intermedius muscle

Sacrum

External iliac arterie

Vastus intermedius muscle

PLATE **72b** MRI SCAN OF PELVIC REGION, FRONTAL SECTION

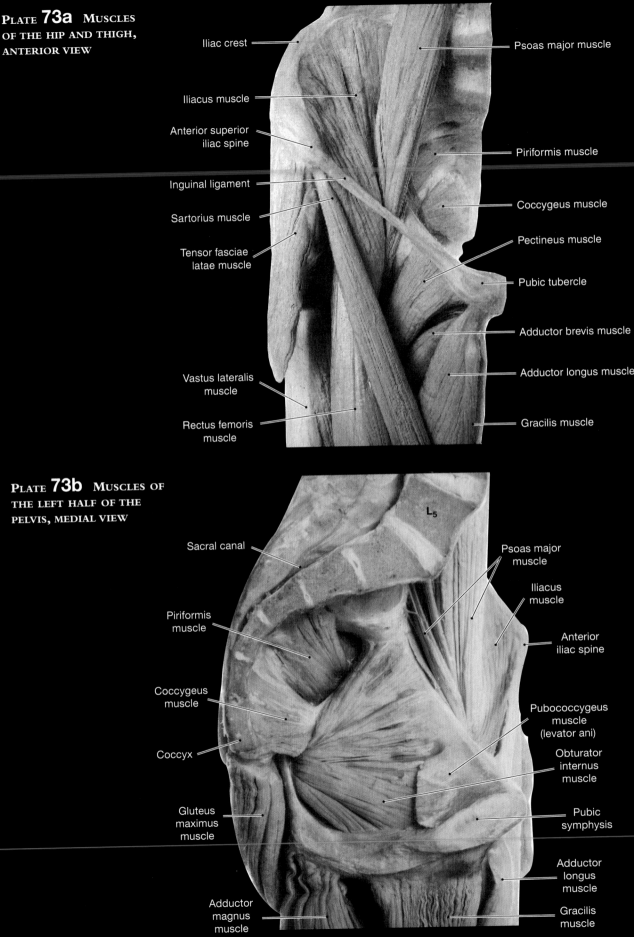

PLATE 73a MUSCLES OF THE HIP AND THIGH, ANTERIOR VIEW

Iliac crest

Iliacus muscle

Anterior superior iliac spine

Inguinal ligament

Sartorius muscle

Tensor fasciae latae muscle

Vastus lateralis muscle

Rectus femoris muscle

Psoas major muscle

Piriformis muscle

Coccygeus muscle

Pectineus muscle

Pubic tubercle

Adductor brevis muscle

Adductor longus muscle

Gracilis muscle

PLATE 73b MUSCLES OF THE LEFT HALF OF THE PELVIS, MEDIAL VIEW

L₅

Sacral canal

Piriformis muscle

Coccygeus muscle

Coccyx

Gluteus maximus muscle

Adductor magnus muscle

Psoas major muscle

Iliacus muscle

Anterior iliac spine

Pubococcygeus muscle (levator ani)

Obturator internus muscle

Pubic symphysis

Adductor longus muscle

Gracilis muscle

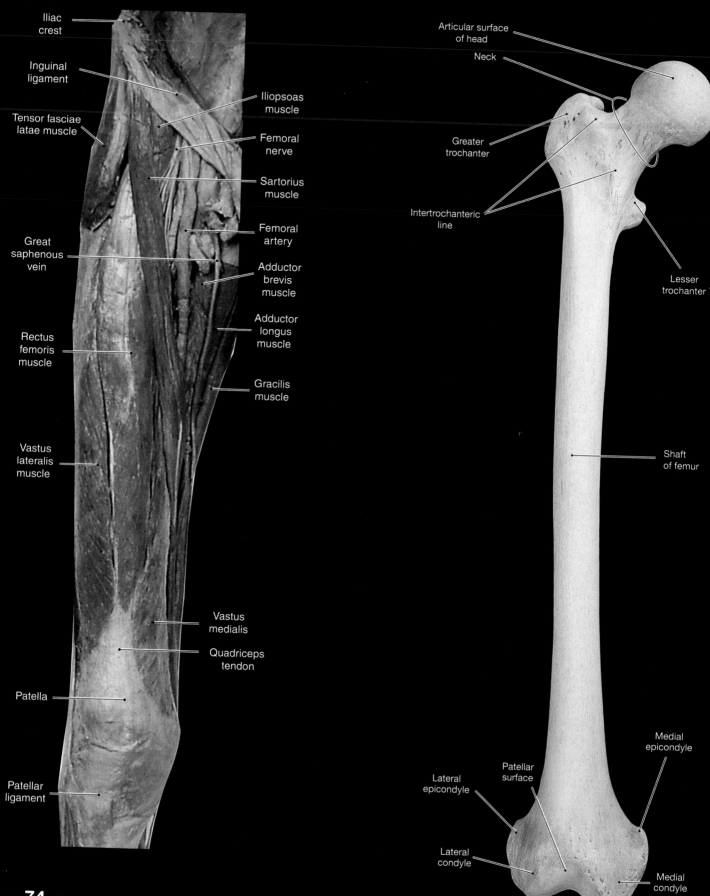

Iliac
crest

Inguinal
ligament

Tensor fasciae
latae muscle

Great
saphenous
vein

Rectus
femoris
muscle

Vastus
lateralis
muscle

Patella

Patellar
ligament

Iliopsoas
muscle

Femoral
nerve

Sartorius
muscle

Femoral
artery

Adductor
brevis
muscle

Adductor
longus
muscle

Gracilis
muscle

Vastus
medialis

Quadriceps
tendon

Articular surface
of head

Neck

Greater
trochanter

Intertrochanteric
line

Lesser
trochanter

Shaft
of femur

Medial
epicondyle

Patellar
surface

Lateral
epicondyle

Lateral
condyle

Medial
condyle

PLATE 74 SUPERFICIAL DISSECTION OF RIGHT LOWER
LIMB, ANTERIOR VIEW

PLATE 75a RIGHT FEMUR, ANTERIOR VIEW

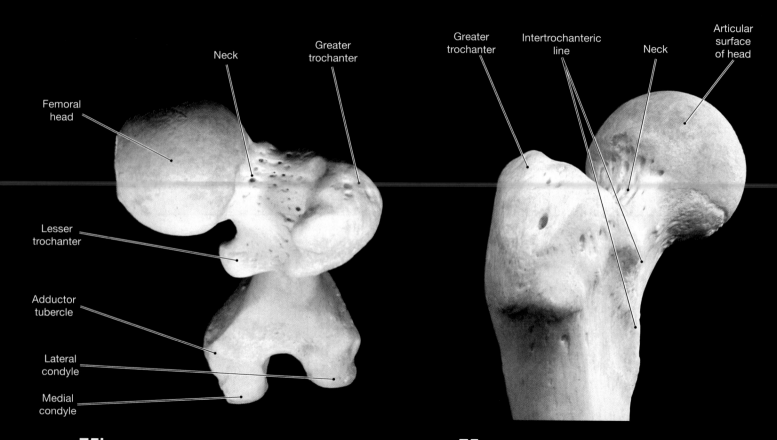

Neck

Greater
trochanter

Femoral
head

Lesser
trochanter

Adductor
tubercle

Lateral
condyle

Medial
condyle

Greater
trochanter

Intertrochanteric
line

Neck

Articular
surface
of head

PLATE **75b** RIGHT FEMUR, SUPERIOR VIEW

PLATE **75c** HEAD OF RIGHT FEMUR, LATERAL VIEW

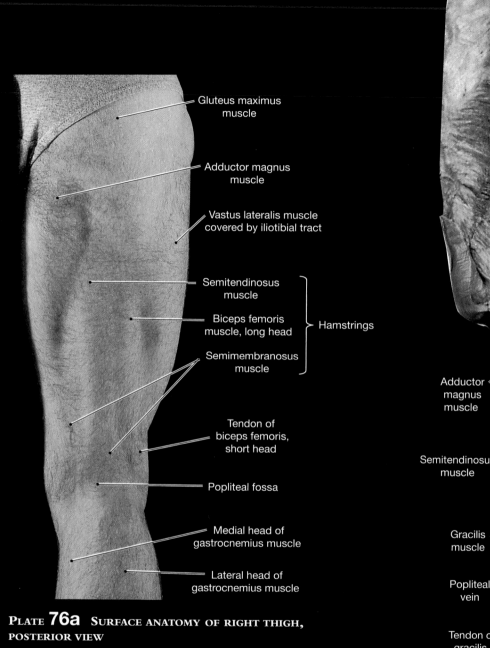

Gluteus maximus muscle

Adductor magnus muscle

Vastus lateralis muscle covered by iliotibial tract

Semitendinosus muscle

Biceps femoris muscle, long head

Semimembranosus muscle

⎤ Hamstrings

Tendon of biceps femoris, short head

Popliteal fossa

Medial head of gastrocnemius muscle

Lateral head of gastrocnemius muscle

PLATE **76a** SURFACE ANATOMY OF RIGHT THIGH, POSTERIOR VIEW

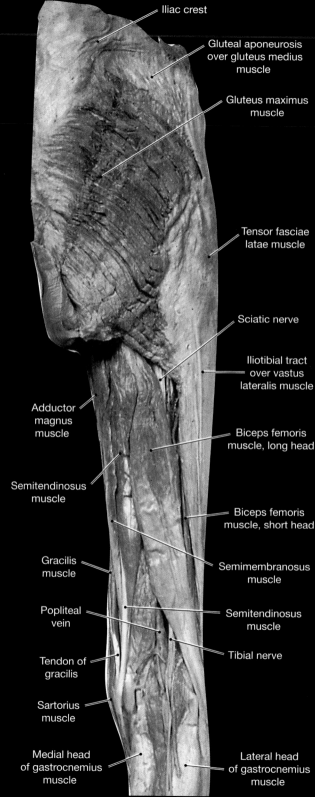

Iliac crest

Gluteal aponeurosis over gluteus medius muscle

Gluteus maximus muscle

Tensor fasciae latae muscle

Sciatic nerve

Iliotibial tract over vastus lateralis muscle

Adductor magnus muscle

Biceps femoris muscle, long head

Semitendinosus muscle

Biceps femoris muscle, short head

Gracilis muscle

Semimembranosus muscle

Popliteal vein

Semitendinosus muscle

Tendon of gracilis

Tibial nerve

Sartorius muscle

Medial head of gastrocnemius muscle

Lateral head of gastrocnemius muscle

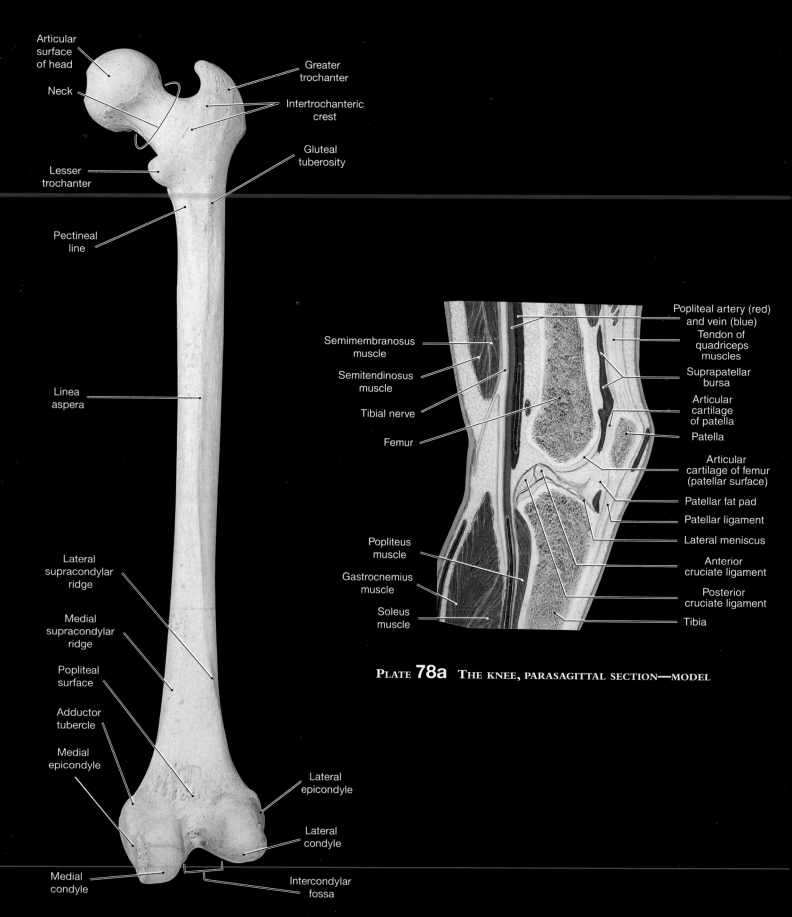

Articular surface of head

Neck

Lesser trochanter

Pectineal line

Linea aspera

Lateral supracondylar ridge

Medial supracondylar ridge

Popliteal surface

Adductor tubercle

Medial epicondyle

Medial condyle

Greater trochanter

Intertrochanteric crest

Gluteal tuberosity

Lateral epicondyle

Lateral condyle

Intercondylar fossa

PLATE 77   **RIGHT FEMUR, POSTERIOR VIEW**

Semimembranosus muscle

Semitendinosus muscle

Tibial nerve

Femur

Popliteus muscle

Gastrocnemius muscle

Soleus muscle

Popliteal artery (red) and vein (blue)

Tendon of quadriceps muscles

Suprapatellar bursa

Articular cartilage of patella

Patella

Articular cartilage of femur (patellar surface)

Patellar fat pad

Patellar ligament

Lateral meniscus

Anterior cruciate ligament

Posterior cruciate ligament

Tibia

PLATE 78a   **THE KNEE, PARASAGITTAL SECTION—MODEL**

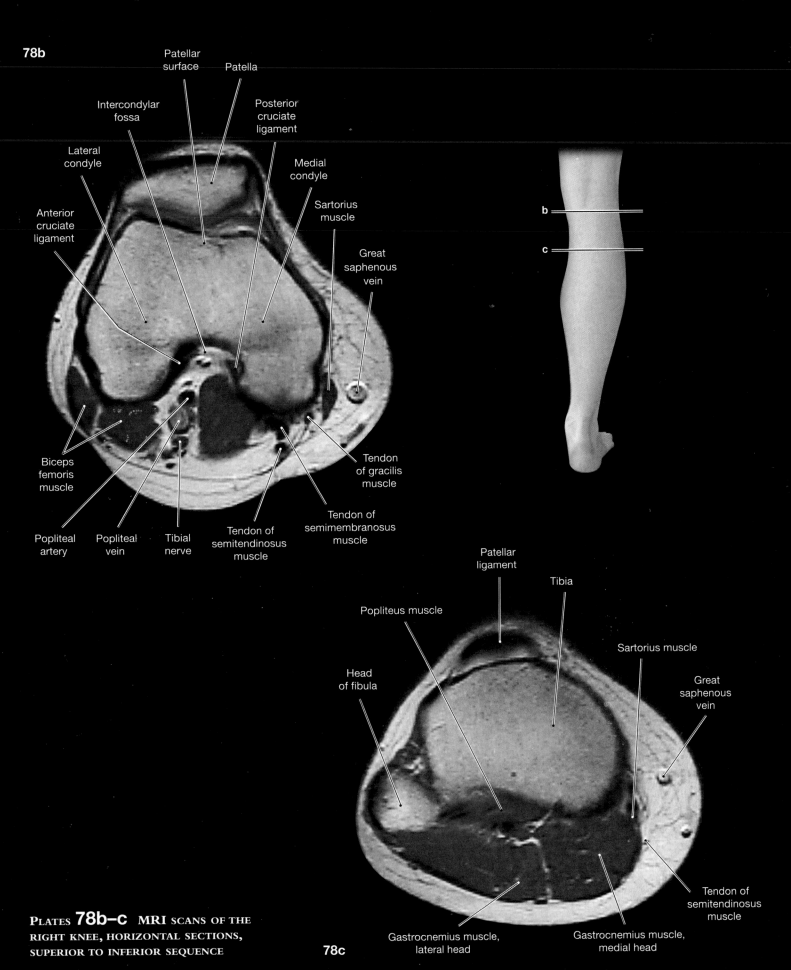

78b

Patellar surface

Patella

Intercondylar fossa

Posterior cruciate ligament

Lateral condyle

Medial condyle

Anterior cruciate ligament

Sartorius muscle

Great saphenous vein

Biceps femoris muscle

Tendon of gracilis muscle

Popliteal artery

Popliteal vein

Tibial nerve

Tendon of semitendinosus muscle

Tendon of semimembranosus muscle

b

c

Patellar ligament

Tibia

Popliteus muscle

Head of fibula

Sartorius muscle

Great saphenous vein

Tendon of semitendinosus muscle

PLATES **78b–c** MRI SCANS OF THE RIGHT KNEE, HORIZONTAL SECTIONS, SUPERIOR TO INFERIOR SEQUENCE

78c

Gastrocnemius muscle, lateral head

Gastrocnemius muscle, medial head

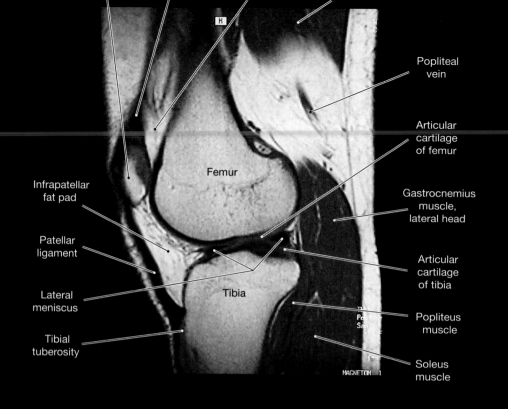

Popliteal
vein

Articular
cartilage
of femur

Femur

Gastrocnemius
muscle,
lateral head

Infrapatellar
fat pad

Patellar
ligament

Articular
cartilage
of tibia

Lateral
meniscus

Tibia

Popliteus
muscle

Tibial
tuberosity

Soleus
muscle

MAGNETOM 1

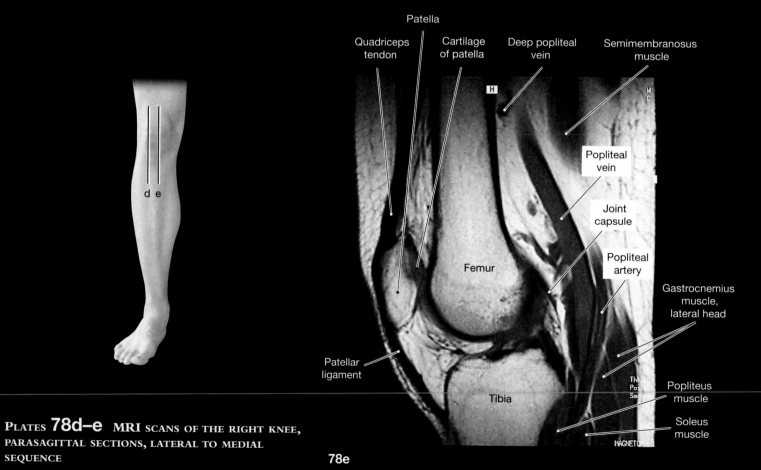

Patella

Quadriceps
tendon

Cartilage
of patella

Deep popliteal
vein

Semimembranosus
muscle

d e

Popliteal
vein

Joint
capsule

Femur

Popliteal
artery

Gastrocnemius
muscle,
lateral head

Patellar
ligament

Tibia

Popliteus
muscle

Soleus
muscle

MAGNETOM

PLATES 78d–e MRI SCANS OF THE RIGHT KNEE,
PARASAGITTAL SECTIONS, LATERAL TO MEDIAL
SEQUENCE

78e

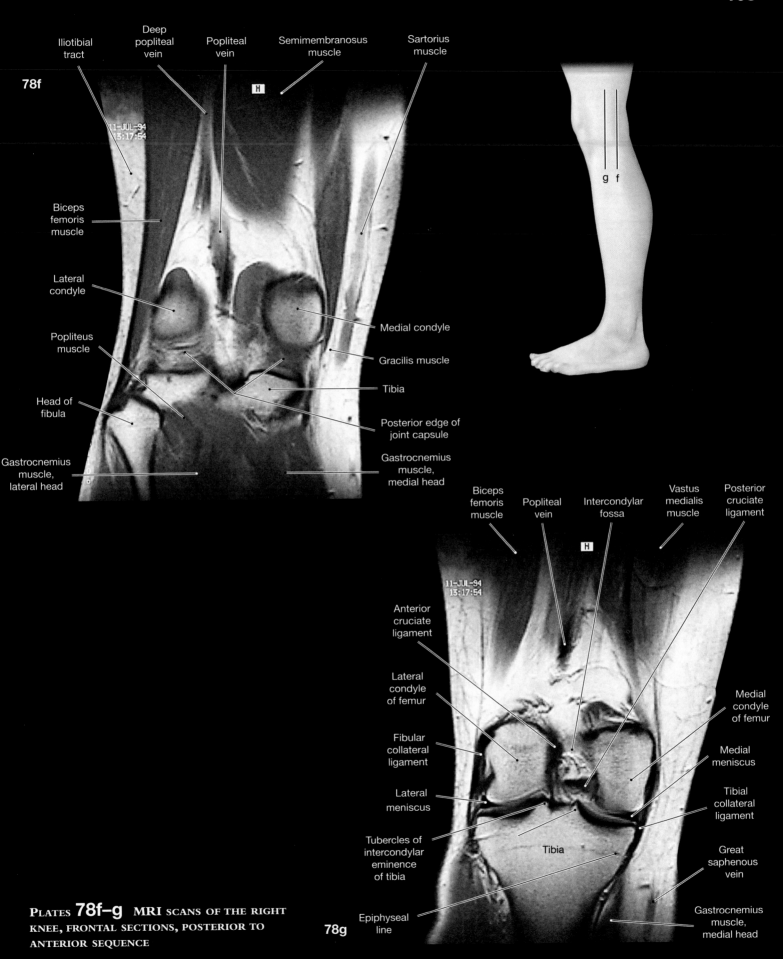

78f

Iliotibial tract
Deep popliteal vein
Popliteal vein
Semimembranosus muscle
Sartorius muscle

Biceps femoris muscle

Lateral condyle

Popliteus muscle

Head of fibula

Gastrocnemius muscle, lateral head

Medial condyle
Gracilis muscle
Tibia
Posterior edge of joint capsule
Gastrocnemius muscle, medial head

g f

Biceps femoris muscle
Popliteal vein
Intercondylar fossa
Vastus medialis muscle
Posterior cruciate ligament

Anterior cruciate ligament

Lateral condyle of femur

Fibular collateral ligament

Lateral meniscus

Tubercles of intercondylar eminence of tibia

Epiphyseal line

Tibia

Medial condyle of femur

Medial meniscus

Tibial collateral ligament

Great saphenous vein

Gastrocnemius muscle, medial head

PLATES 78f–g MRI SCANS OF THE RIGHT KNEE, FRONTAL SECTIONS, POSTERIOR TO ANTERIOR SEQUENCE

78g

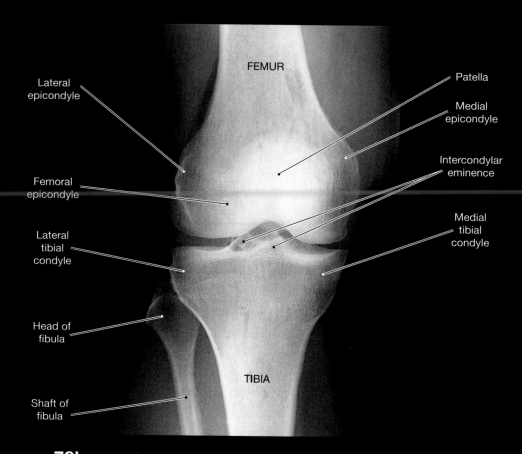

Lateral
epicondyle

FEMUR

Patella

Medial
epicondyle

Intercondylar
eminence

Femoral
epicondyle

Lateral
tibial
condyle

Medial
tibial
condyle

Head of
fibula

TIBIA

Shaft of
fibula

PLATE **78h** X-RAY OF THE EXTENDED RIGHT KNEE, ANTERIOR–POSTERIOR PROJECTION

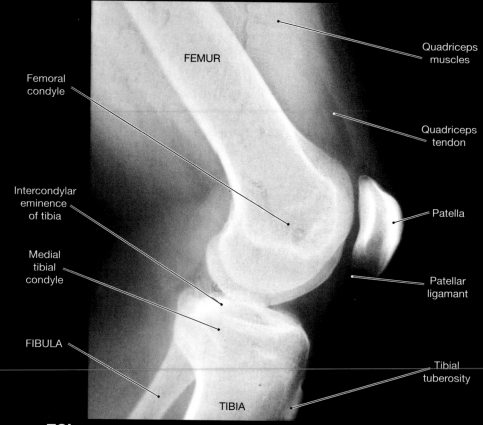

FEMUR

Quadriceps
muscles

Femoral
condyle

Quadriceps
tendon

Intercondylar
eminence
of tibia

Patella

Medial
tibial
condyle

Patellar
ligamant

FIBULA

Tibial
tuberosity

TIBIA

PLATE **78i** X-RAY OF THE PARTIALLY FLEXED RIGHT KNEE, LATERAL PROJECTION

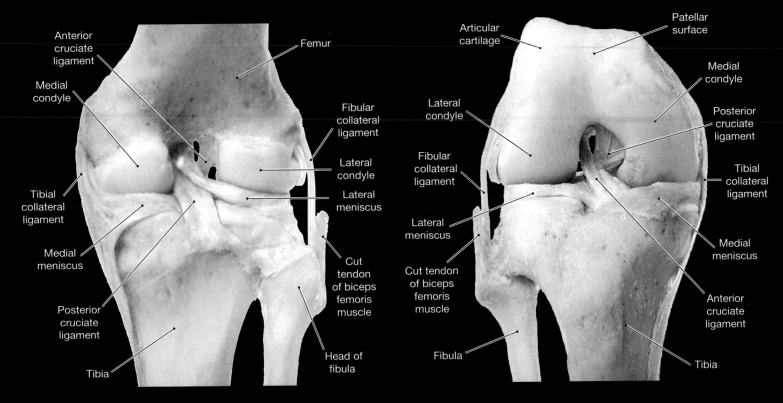

Anterior cruciate ligament

Medial condyle

Tibial collateral ligament

Medial meniscus

Posterior cruciate ligament

Tibia

Femur

Fibular collateral ligament

Lateral condyle

Lateral meniscus

Cut tendon of biceps femoris muscle

Head of fibula

PLATE 79a EXTENDED RIGHT KNEE, POSTERIOR VIEW

Articular cartilage

Lateral condyle

Fibular collateral ligament

Lateral meniscus

Cut tendon of biceps femoris muscle

Fibula

Patellar surface

Medial condyle

Posterior cruciate ligament

Tibial collateral ligament

Medial meniscus

Anterior cruciate ligament

Tibia

PLATE 79b EXTENDED RIGHT KNEE, ANTERIOR VIEW

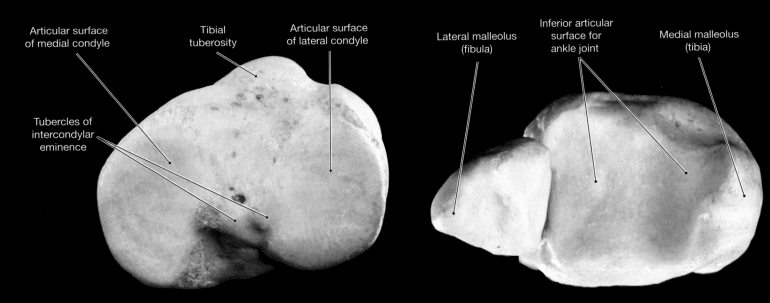

Articular surface of medial condyle

Tibial tuberosity

Articular surface of lateral condyle

Tubercles of intercondylar eminence

PLATE 80a PROXIMAL END OF RIGHT TIBIA, SUPERIOR VIEW

Lateral malleolus (fibula)

Inferior articular surface for ankle joint

Medial malleolus (tibia)

PLATE 80b DISTAL END OF TIBIA AND FIBULA, INFERIOR VIEW

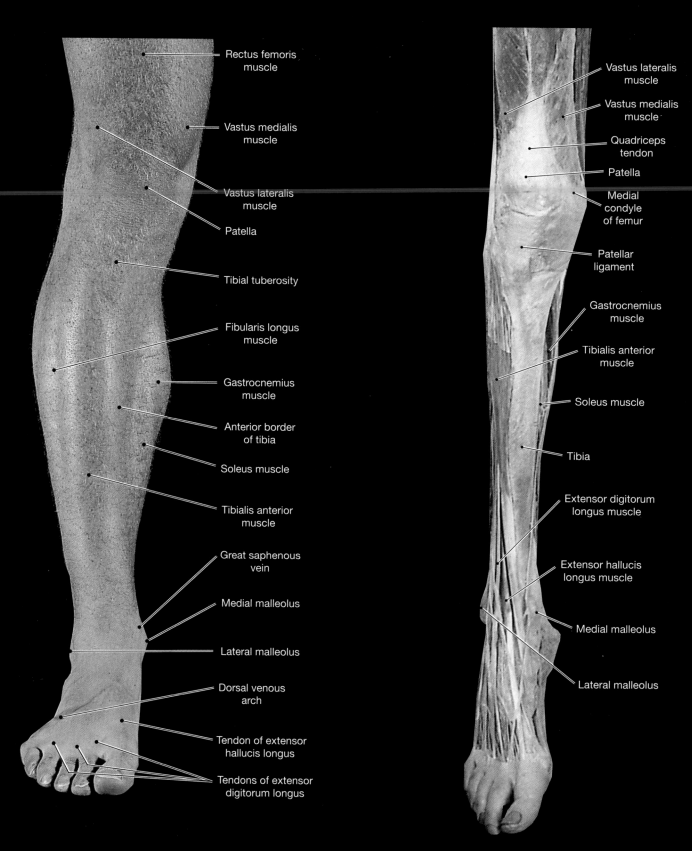

Rectus femoris muscle

Vastus medialis muscle

Vastus lateralis muscle

Patella

Tibial tuberosity

Fibularis longus muscle

Gastrocnemius muscle

Anterior border of tibia

Soleus muscle

Tibialis anterior muscle

Great saphenous vein

Medial malleolus

Lateral malleolus

Dorsal venous arch

Tendon of extensor hallucis longus

Tendons of extensor digitorum longus

Vastus lateralis muscle

Vastus medialis muscle

Quadriceps tendon

Patella

Medial condyle of femur

Patellar ligament

Gastrocnemius muscle

Tibialis anterior muscle

Soleus muscle

Tibia

Extensor digitorum longus muscle

Extensor hallucis longus muscle

Medial malleolus

Lateral malleolus

PLATE **81a** SURFACE ANATOMY OF THE RIGHT LEG AND FOOT, ANTERIOR VIEW

PLATE **81b** SUPERFICIAL DISSECTION OF THE RIGHT LEG AND FOOT, ANTERIOR VIEW

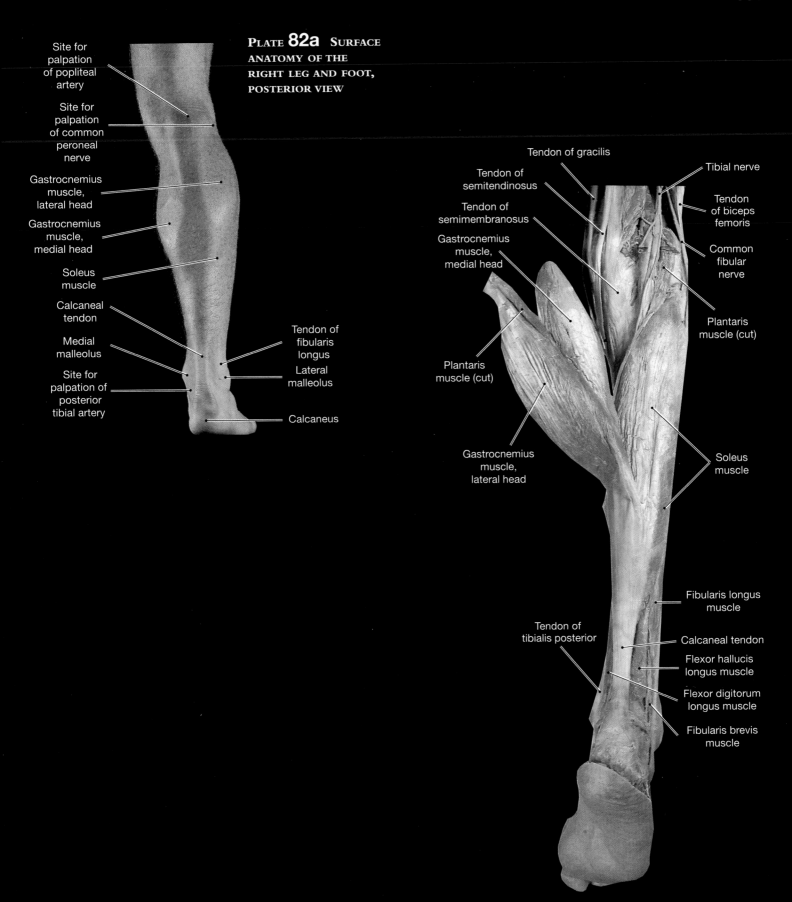

Site for palpation of popliteal artery

Site for palpation of common peroneal nerve

Gastrocnemius muscle, lateral head

Gastrocnemius muscle, medial head

Soleus muscle

Calcaneal tendon

Medial malleolus

Site for palpation of posterior tibial artery

Tendon of fibularis longus

Lateral malleolus

Calcaneus

PLATE **82a** SURFACE ANATOMY OF THE RIGHT LEG AND FOOT, POSTERIOR VIEW

Tendon of gracilis

Tendon of semitendinosus

Tendon of semimembranosus

Gastrocnemius muscle, medial head

Plantaris muscle (cut)

Gastrocnemius muscle, lateral head

Tibial nerve

Tendon of biceps femoris

Common fibular nerve

Plantaris muscle (cut)

Soleus muscle

Fibularis longus muscle

Tendon of tibialis posterior

Calcaneal tendon

Flexor hallucis longus muscle

Flexor digitorum longus muscle

Fibularis brevis muscle

PLATE **82b** SUPERFICIAL DISSECTION OF THE RIGHT LEG AND FOOT, POSTERIOR VIEW

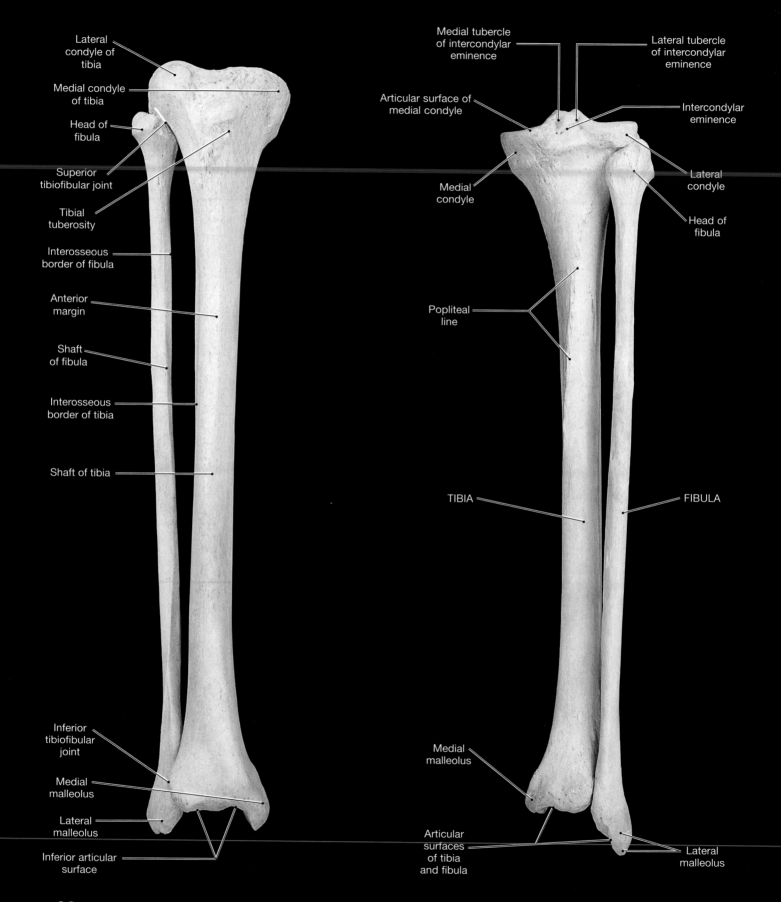

Lateral condyle of tibia

Medial condyle of tibia

Head of fibula

Superior tibiofibular joint

Tibial tuberosity

Interosseous border of fibula

Anterior margin

Shaft of fibula

Interosseous border of tibia

Shaft of tibia

Inferior tibiofibular joint

Medial malleolus

Lateral malleolus

Inferior articular surface

Medial tubercle of intercondylar eminence

Lateral tubercle of intercondylar eminence

Articular surface of medial condyle

Intercondylar eminence

Medial condyle

Lateral condyle

Head of fibula

Popliteal line

TIBIA

FIBULA

Medial malleolus

Articular surfaces of tibia and fibula

Lateral malleolus

PLATE **83a**   TIBIA AND FIBULA, ANTERIOR VIEW

PLATE **83b**   TIBIA AND FIBULA, POSTERIOR VIEW

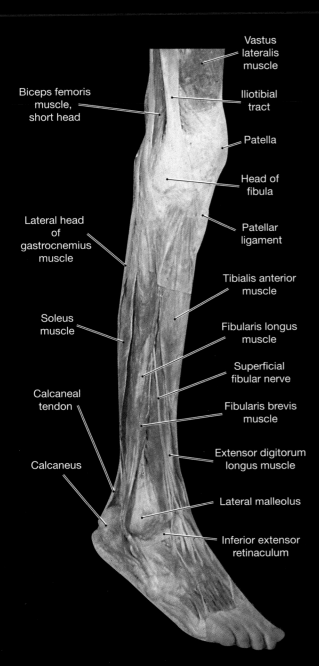

Vastus
lateralis
muscle

Biceps femoris
muscle,
short head

Iliotibial
tract

Patella

Head of
fibula

Lateral head
of
gastrocnemius
muscle

Patellar
ligament

Tibialis anterior
muscle

Soleus
muscle

Fibularis longus
muscle

Superficial
fibular nerve

Calcaneal
tendon

Fibularis brevis
muscle

Calcaneus

Extensor digitorum
longus muscle

Lateral malleolus

Inferior extensor
retinaculum

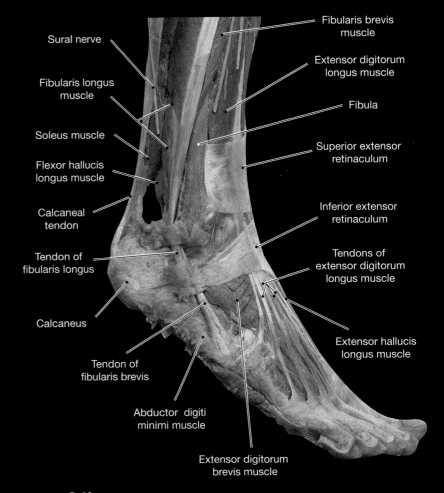

Sural nerve

Fibularis longus
muscle

Soleus muscle

Flexor hallucis
longus muscle

Calcaneal
tendon

Tendon of
fibularis longus

Calcaneus

Tendon of
fibularis brevis

Abductor digiti
minimi muscle

Extensor digitorum
brevis muscle

Fibularis brevis
muscle

Extensor digitorum
longus muscle

Fibula

Superior extensor
retinaculum

Inferior extensor
retinaculum

Tendons of
extensor digitorum
longus muscle

Extensor hallucis
longus muscle

PLATE 84a SUPERFICIAL DISSECTION OF THE
RIGHT LEG AND FOOT, ANTEROLATERAL VIEW

PLATE 84b SUPERFICIAL DISSECTION OF THE RIGHT FOOT,
LATERAL VIEW

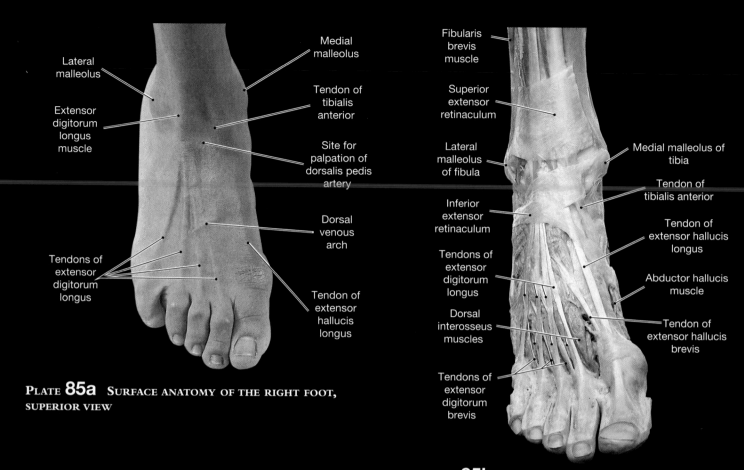

Lateral malleolus

Extensor digitorum longus muscle

Tendons of extensor digitorum longus

Medial malleolus

Tendon of tibialis anterior

Site for palpation of dorsalis pedis artery

Dorsal venous arch

Tendon of extensor hallucis longus

PLATE 85a SURFACE ANATOMY OF THE RIGHT FOOT, SUPERIOR VIEW

Fibularis brevis muscle

Superior extensor retinaculum

Lateral malleolus of fibula

Inferior extensor retinaculum

Tendons of extensor digitorum longus

Dorsal interosseus muscles

Tendons of extensor digitorum brevis

Medial malleolus of tibia

Tendon of tibialis anterior

Tendon of extensor hallucis longus

Abductor hallucis muscle

Tendon of extensor hallucis brevis

PLATE 85b SUPERFICIAL DISSECTION OF THE RIGHT FOOT, SUPERIOR VIEW

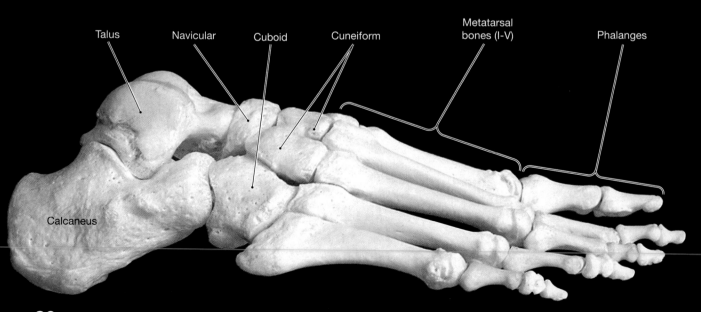

Talus   Navicular   Cuboid   Cuneiform   Metatarsal bones (I-V)   Phalanges

Calcaneus

PLATE 86a BONES OF THE RIGHT FOOT, LATERAL VIEW

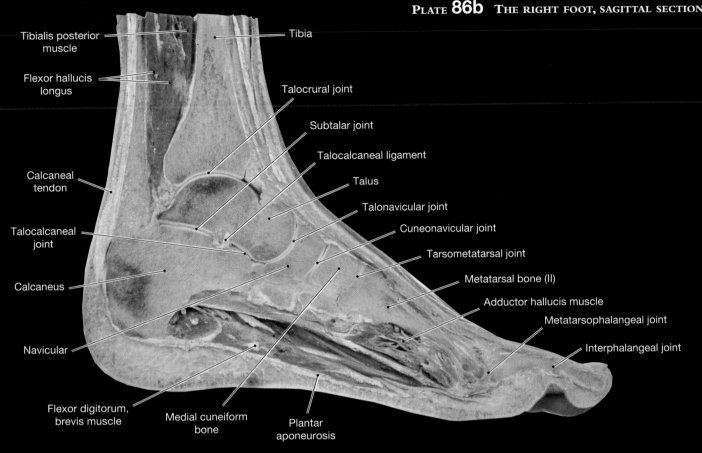

Tibialis posterior muscle

Flexor hallucis longus

Calcaneal tendon

Talocalcaneal joint

Calcaneus

Navicular

Flexor digitorum, brevis muscle

Medial cuneiform bone

Plantar aponeurosis

Tibia

Talocrural joint

Subtalar joint

Talocalcaneal ligament

Talus

Talonavicular joint

Cuneonavicular joint

Tarsometatarsal joint

Metatarsal bone (II)

Adductor hallucis muscle

Metatarsophalangeal joint

Interphalangeal joint

PLATE **86c** MRI SCAN OF THE RIGHT ANKLE, SAGITTAL SECTION

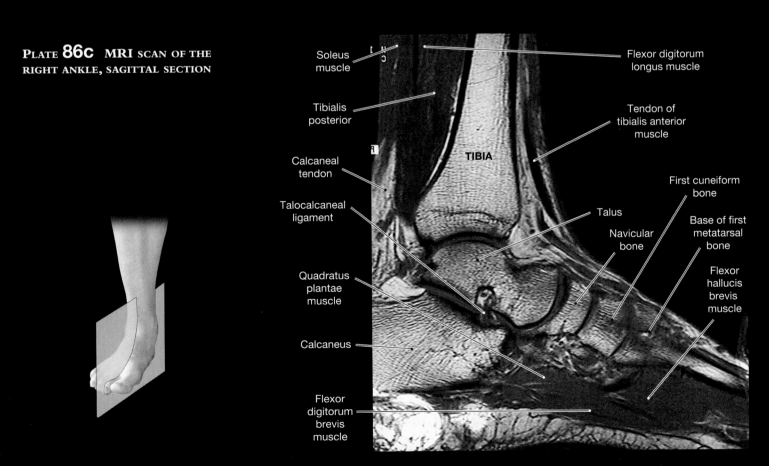

Soleus muscle

Tibialis posterior

Calcaneal tendon

Talocalcaneal ligament

Quadratus plantae muscle

Calcaneus

Flexor digitorum brevis muscle

Flexor digitorum longus muscle

Tendon of tibialis anterior muscle

TIBIA

First cuneiform bone

Talus

Base of first metatarsal bone

Navicular bone

Flexor hallucis brevis muscle

PLATE **87a** ANKLE AND FOOT, POSTERIOR VIEW

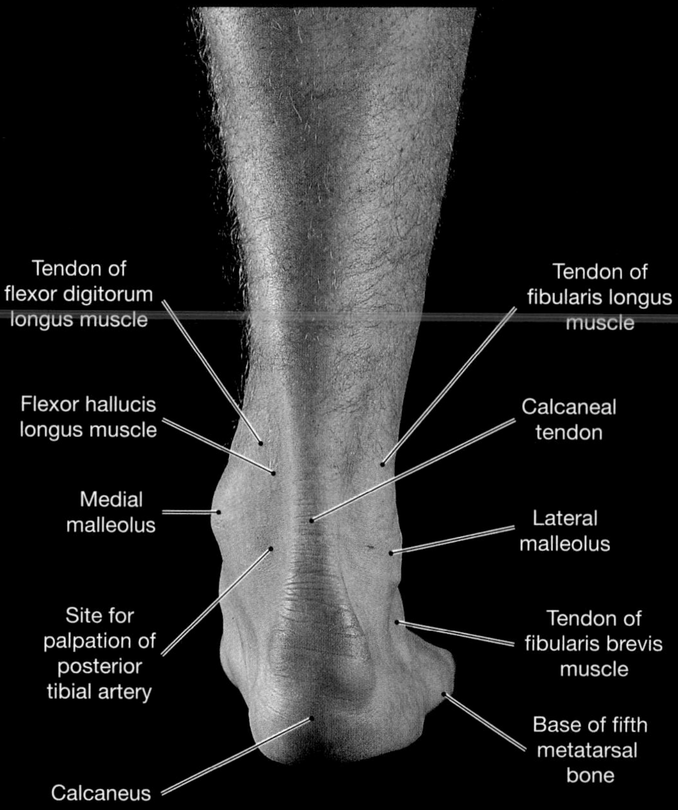

Tendon of flexor digitorum longus muscle

Tendon of fibularis longus muscle

Flexor hallucis longus muscle

Calcaneal tendon

Medial malleolus

Lateral malleolus

Site for palpation of posterior tibial artery

Tendon of fibularis brevis muscle

Base of fifth metatarsal bone

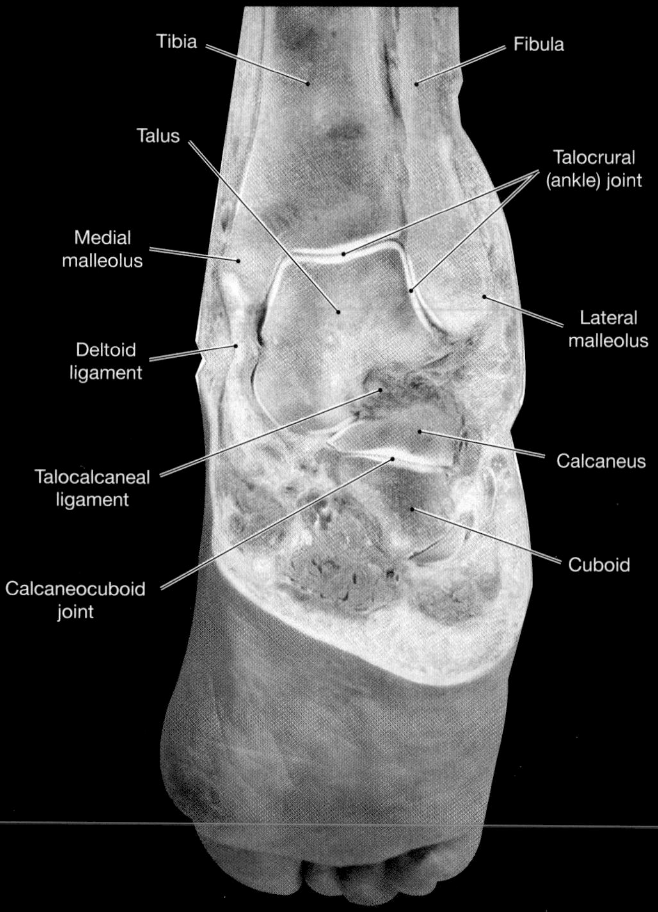

Calcaneus

Tibia

Fibula

Talus

Talocrural (ankle) joint

Medial malleolus

Deltoid ligament

Lateral malleolus

Talocalcaneal ligament

Calcaneus

Calcaneocuboid joint

Cuboid

PLATE **87b** FRONTAL SECTION THROUGH THE RIGHT FOOT, POSTERIOR VIEW

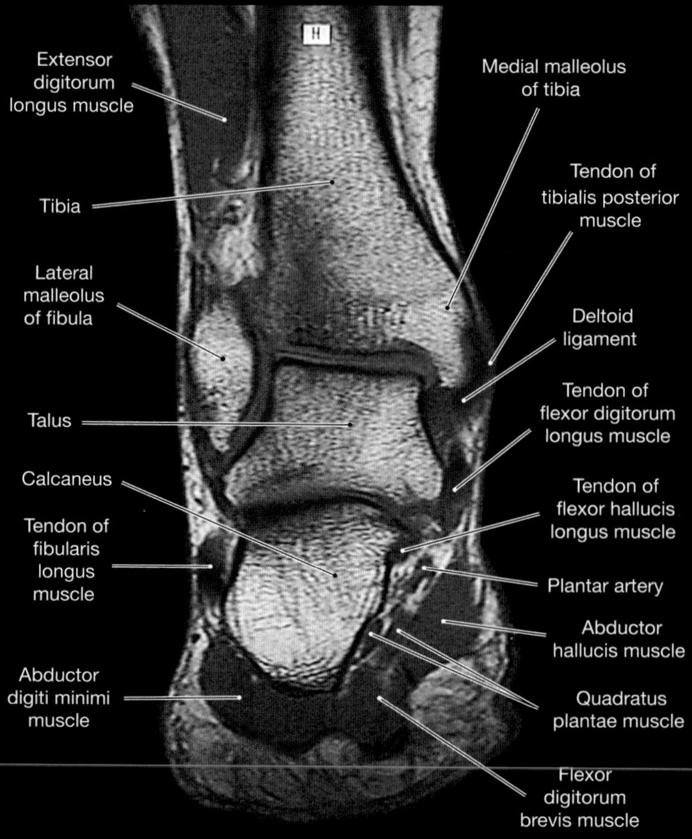

Extensor digitorum longus muscle

Medial malleolus of tibia

Tibia

Tendon of tibialis posterior muscle

Lateral malleolus of fibula

Deltoid ligament

Talus

Tendon of flexor digitorum longus muscle

Calcaneus

Tendon of flexor hallucis longus muscle

Tendon of fibularis longus muscle

Plantar artery

Abductor hallucis muscle

Abductor digiti minimi muscle

Quadratus plantae muscle

Flexor digitorum brevis muscle

PLATE **87c** MRI SCAN OF THE RIGHT ANKLE, FRONTAL SECTION

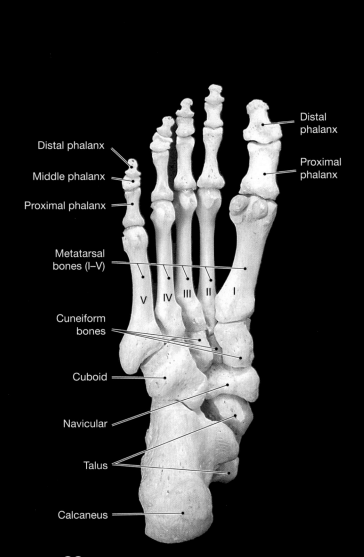

Distal phalanx

Distal phalanx

Middle phalanx

Proximal phalanx

Proximal phalanx

Distal phalanx

Proximal phalanx

Metatarsal bones (I–V)

V   IV   III   II   I

Cuneiform bones

Cuboid

Navicular

Talus

Calcaneus

PLATE 88 BONES OF THE RIGHT FOOT, INFERIOR (PLANTAR) VIEW

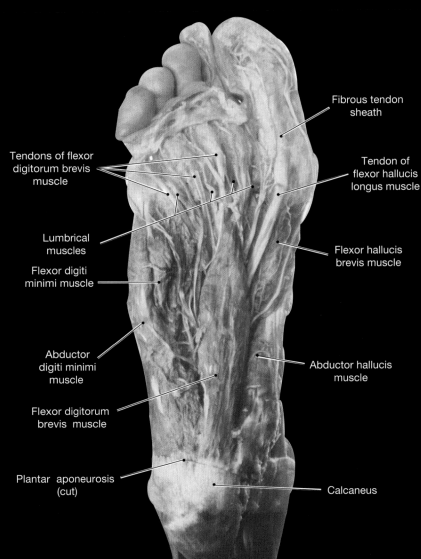

Fibrous tendon sheath

Tendons of flexor digitorum brevis muscle

Tendon of flexor hallucis longus muscle

Lumbrical muscles

Flexor digiti minimi muscle

Flexor hallucis brevis muscle

Abductor digiti minimi muscle

Abductor hallucis muscle

Flexor digitorum brevis muscle

Plantar aponeurosis (cut)

Calcaneus

PLATE 89 SUPERFICIAL DISSECTION OF THE RIGHT FOOT, PLANTAR VIEW

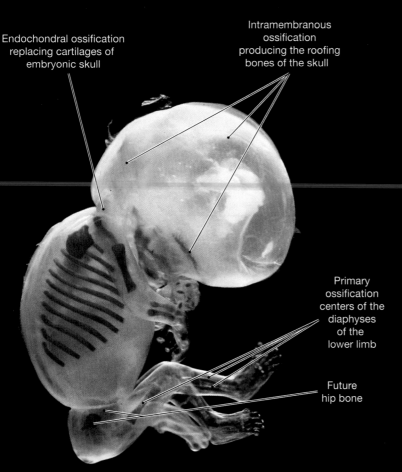

Endochondral ossification replacing cartilages of embryonic skull

Intramembranous ossification producing the roofing bones of the skull

Primary ossification centers of the diaphyses of the lower limb

Future hip bone

PLATE **90a**   SKELETON OF FETUS AFTER **10 WEEKS OF** DEVELOPMENT

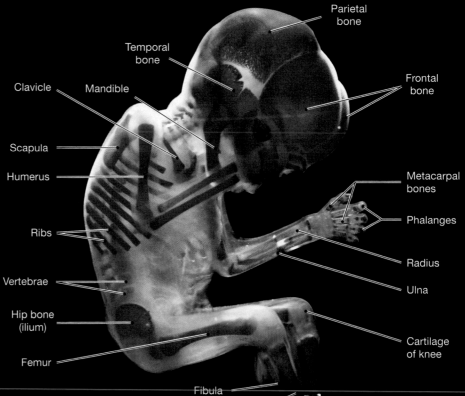

Parietal bone

Temporal bone

Frontal bone

Clavicle   Mandible

Scapula

Humerus

Metacarpal bones

Phalanges

Ribs

Radius

Vertebrae

Ulna

Hip bone (ilium)

Cartilage of knee

Femur

Fibula

Tibia

Phalanx

Metatarsal bones

PLATE **90b**   SKELETON OF FETUS AFTER **16** WEEKS OF DEVELOPMENT

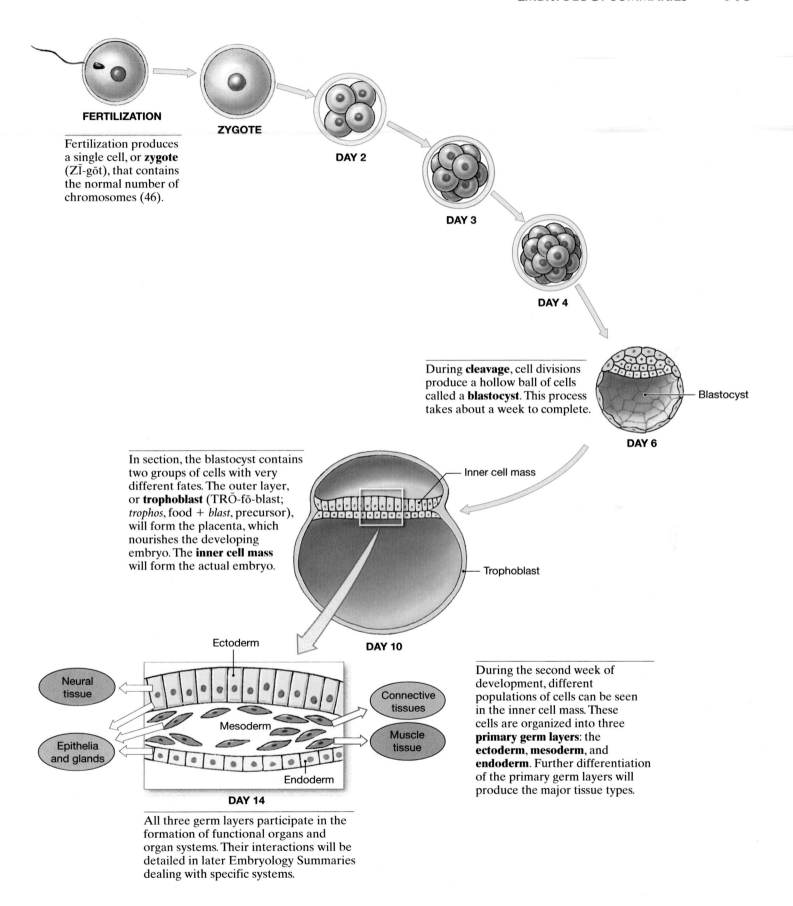

FERTILIZATION

ZYGOTE

Fertilization produces a single cell, or **zygote** (ZĪ-gōt), that contains the normal number of chromosomes (46).

DAY 2

DAY 3

DAY 4

During **cleavage**, cell divisions produce a hollow ball of cells called a **blastocyst**. This process takes about a week to complete.

Blastocyst

DAY 6

In section, the blastocyst contains two groups of cells with very different fates. The outer layer, or **trophoblast** (TRŌ-fō-blast; *trophos*, food + *blast*, precursor), will form the placenta, which nourishes the developing embryo. The **inner cell mass** will form the actual embryo.

Inner cell mass

Trophoblast

DAY 10

Ectoderm

Neural tissue

Epithelia and glands

Mesoderm

Endoderm

Connective tissues

Muscle tissue

During the second week of development, different populations of cells can be seen in the inner cell mass. These cells are organized into three **primary germ layers**: the **ectoderm**, **mesoderm**, and **endoderm**. Further differentiation of the primary germ layers will produce the major tissue types.

DAY 14

All three germ layers participate in the formation of functional organs and organ systems. Their interactions will be detailed in later Embryology Summaries dealing with specific systems.

EMBRYOLOGY SUMMARY **1:** THE FORMATION OF TISSUES

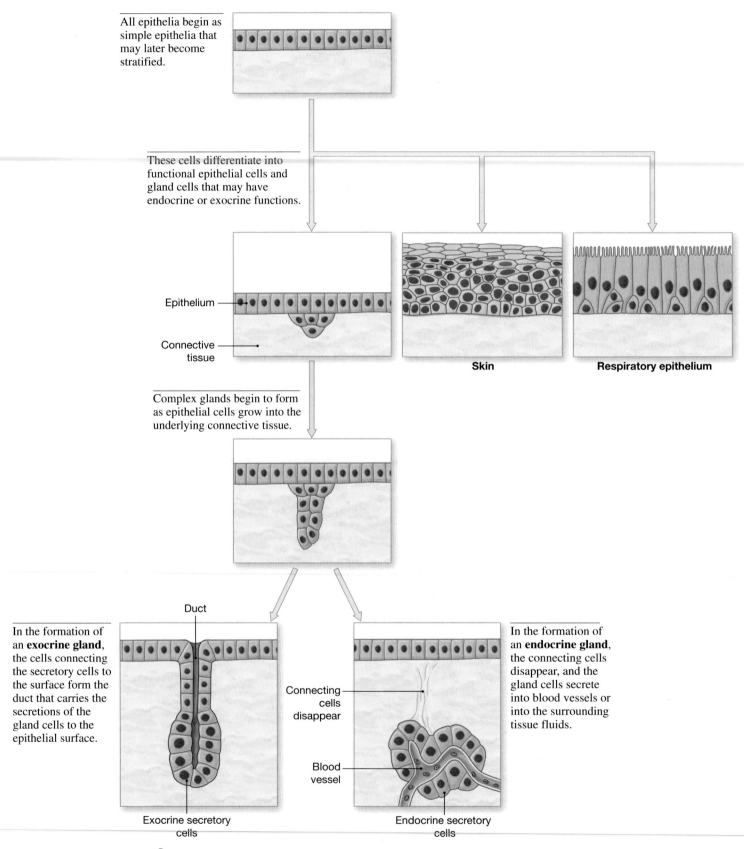

All epithelia begin as simple epithelia that may later become stratified.

These cells differentiate into functional epithelial cells and gland cells that may have endocrine or exocrine functions.

Epithelium

Connective tissue

Skin

Respiratory epithelium

Complex glands begin to form as epithelial cells grow into the underlying connective tissue.

Duct

In the formation of an **exocrine gland**, the cells connecting the secretory cells to the surface form the duct that carries the secretions of the gland cells to the epithelial surface.

Connecting cells disappear

In the formation of an **endocrine gland**, the connecting cells disappear, and the gland cells secrete into blood vessels or into the surrounding tissue fluids.

Blood vessel

Exocrine secretory cells

Endocrine secretory cells

EMBRYOLOGY SUMMARY 2: THE DEVELOPMENT OF EPITHELIA

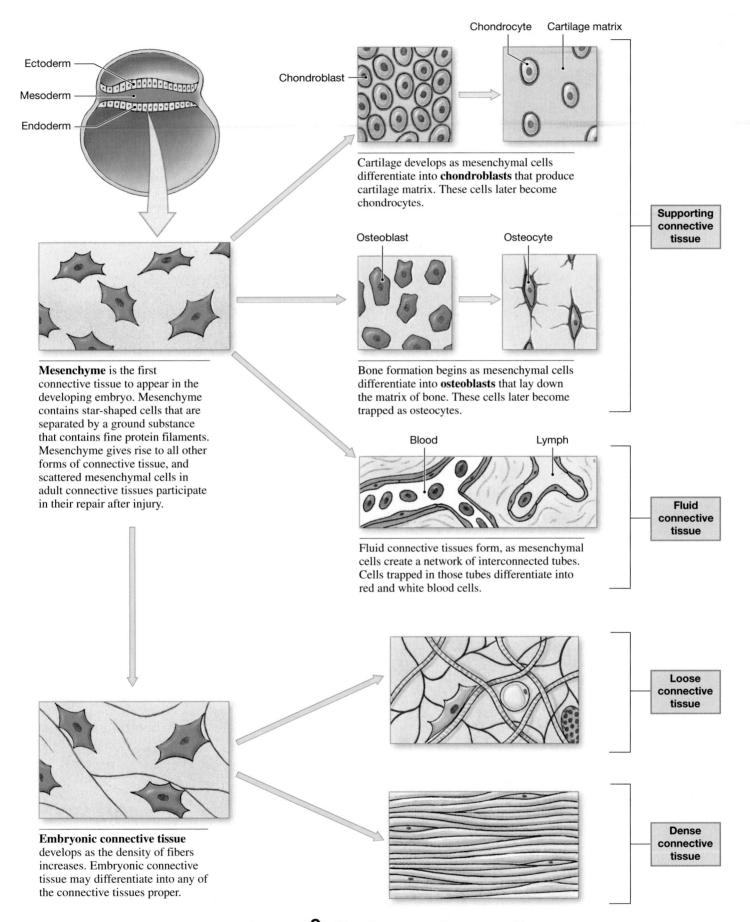

Ectoderm
Mesoderm
Endoderm

Chondroblast
Chondrocyte   Cartilage matrix

Cartilage develops as mesenchymal cells differentiate into **chondroblasts** that produce cartilage matrix. These cells later become chondrocytes.

Osteoblast
Osteocyte

Bone formation begins as mesenchymal cells differentiate into **osteoblasts** that lay down the matrix of bone. These cells later become trapped as osteocytes.

Blood
Lymph

Fluid connective tissues form, as mesenchymal cells create a network of interconnected tubes. Cells trapped in those tubes differentiate into red and white blood cells.

Mesenchyme is the first connective tissue to appear in the developing embryo. Mesenchyme contains star-shaped cells that are separated by a ground substance that contains fine protein filaments. Mesenchyme gives rise to all other forms of connective tissue, and scattered mesenchymal cells in adult connective tissues participate in their repair after injury.

Embryonic connective tissue develops as the density of fibers increases. Embryonic connective tissue may differentiate into any of the connective tissues proper.

Supporting connective tissue

Fluid connective tissue

Loose connective tissue

Dense connective tissue

EMBRYOLOGY SUMMARY **3:** THE ORIGINS OF CONNECTIVE TISSUES

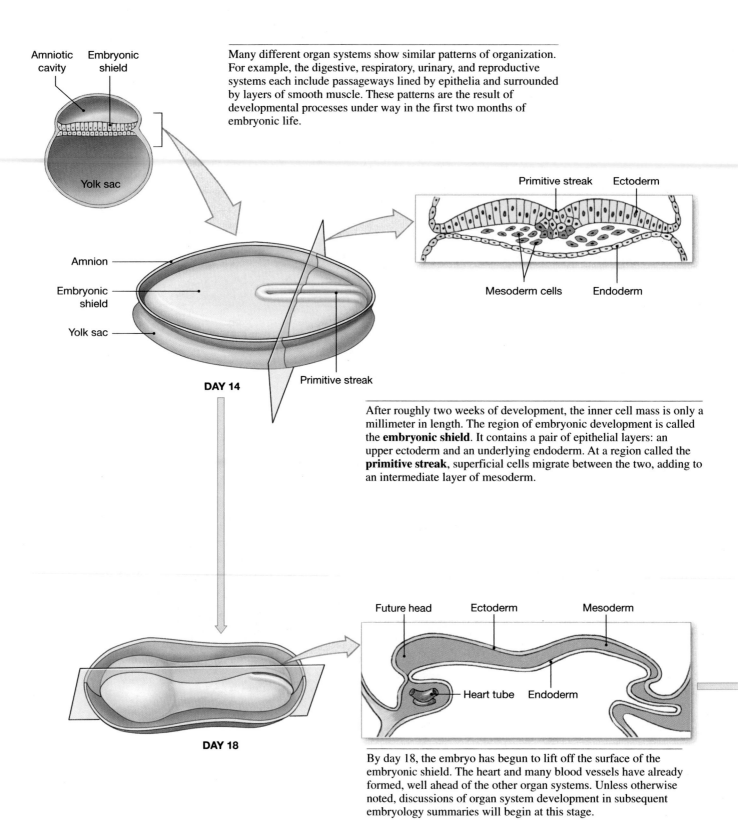

Amniotic cavity

Embryonic shield

Yolk sac

Many different organ systems show similar patterns of organization. For example, the digestive, respiratory, urinary, and reproductive systems each include passageways lined by epithelia and surrounded by layers of smooth muscle. These patterns are the result of developmental processes under way in the first two months of embryonic life.

Primitive streak　　Ectoderm

Mesoderm cells　　Endoderm

Amnion

Embryonic shield

Yolk sac

DAY 14　　Primitive streak

After roughly two weeks of development, the inner cell mass is only a millimeter in length. The region of embryonic development is called the **embryonic shield**. It contains a pair of epithelial layers: an upper ectoderm and an underlying endoderm. At a region called the **primitive streak**, superficial cells migrate between the two, adding to an intermediate layer of mesoderm.

Future head　　Ectoderm　　Mesoderm

Heart tube　　Endoderm

DAY 18

By day 18, the embryo has begun to lift off the surface of the embryonic shield. The heart and many blood vessels have already formed, well ahead of the other organ systems. Unless otherwise noted, discussions of organ system development in subsequent embryology summaries will begin at this stage.

DERIVATIVES OF PRIMARY GERM LAYERS

Ectoderm Forms: Epidermis and epidermal derivatives of the integumentary system, including hair follicles, nails, and
glands communicating with the skin surface (sweat, milk, and sebum)
Lining of the mouth, salivary glands, nasal passageways, and anus
Nervous system, including brain and spinal cord
Portions of endocrine system (pituitary gland and parts of suprarenal glands)
Portions of skull, pharyngeal arches, and teeth

Mesoderm Forms: Dermis of integumentary system
Lining of the body cavities (pleural, pericardial, peritoneal)
Muscular, skeletal, cardiovascular, and lymphoid systems
Kidneys and part of the urinary tract
Gonads and most of the reproductive tract
Connective tissues supporting all organ systems
Portions of endocrine system (parts of suprarenal glands and endocrine tissues of the reproductive tract)

Endoderm Forms: Most of the digestive system: epithelium (except mouth and anus), exocrine glands (except
salivary glands), the liver and pancreas
Most of the respiratory system: epithelium (except nasal passageways) and mucous glands
Portions of urinary and reproductive systems (ducts and the stem cells that produce gametes)
Portions of endocrine system (thymus, thyroid gland, parathyroid glands, and pancreas)

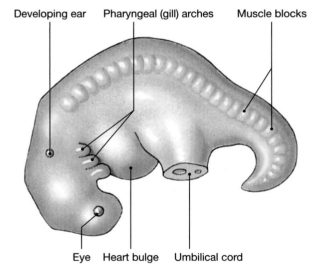

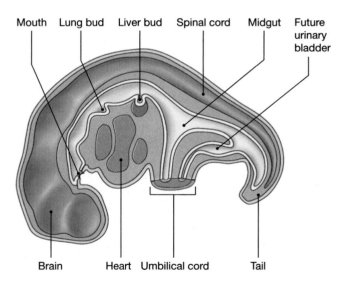

Developing ear   Pharyngeal (gill) arches   Muscle blocks

Eye   Heart bulge   Umbilical cord

Mouth   Lung bud   Liver bud   Spinal cord   Midgut   Future urinary bladder

Brain   Heart   Umbilical cord   Tail

DAY 28

After one month, you can find the beginnings of all major organ systems.
The role of each of the primary germ layers in the formation of organs is
summarized in the accompanying table; details are given in later
Embryology Summaries.

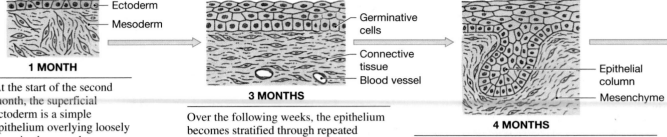

1 MONTH

At the start of the second month, the superficial ectoderm is a simple epithelium overlying loosely organized mesenchyme.

3 MONTHS

- Germinative cells
- Connective tissue
- Blood vessel

Over the following weeks, the epithelium becomes stratified through repeated divisions of the *basal* or *germinative cells*. The underlying mesenchyme differentiates into embryonic connective tissue containing blood vessels that bring nutrients to the region.

4 MONTHS

- Epithelial column
- Mesenchyme

During the third and fourth months, small areas of epidermis undergo extensive divisions and form cords of cells that grow into the dermis. These are **epithelial columns**. Mesenchymal cells surround the columns as they extend deeper and deeper into the dermis. Hair follicles, sebaceous glands, and sweat glands develop from these columns.

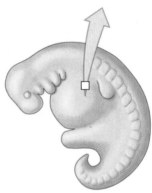

NAILS

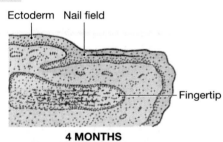

- Ectoderm
- Nail field
- Fingertip

4 MONTHS

Nails begin as thickenings of the epidermis near the tips of the fingers and toes. These thickenings settle into the dermis, and the borderline with the general epidermis becomes distinct. Initially, nail production involves all of the germinative cells of the *nail field*.

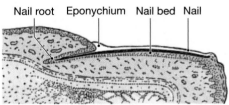

- Nail root
- Eponychium
- Nail bed
- Nail

BIRTH

By the time of birth, nail production is restricted to the *nail root*.

SKIN

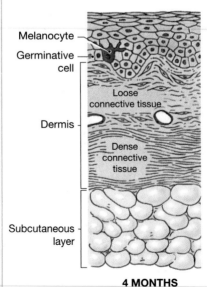

- Melanocyte
- Germinative cell
- Loose connective tissue
- Dermis
- Dense connective tissue
- Subcutaneous layer

4 MONTHS

As basal cell divisions continue, the epithelial layer thickens and the basal lamina is thrown into irregular folds. Pigment cells called *melanocytes* migrate into the area and squeeze between the germinative cells. The epithelium now resembles the *epidermis* of the adult.

The embryonic connective tissue differentiates into the *dermis*. Fibroblasts and other connective tissue cells form from mesenchymal cells or migrate into the area. The density of fibers increases. Loose connective tissue extends into the ridges, but a deeper, less vascular region is dominated by a dense, irregular collagen fiber network. Below the dermis the embryonic connective tissue develops into the *subcutaneous layer*, a layer of loose connective tissue.

EMBRYOLOGY SUMMARY 5: THE DEVELOPMENT OF THE INTEGUMENTARY SYSTEM

HAIR FOLLICLES

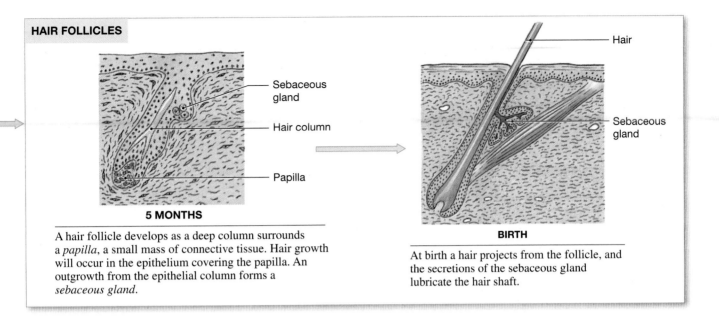

5 MONTHS

A hair follicle develops as a deep column surrounds a *papilla*, a small mass of connective tissue. Hair growth will occur in the epithelium covering the papilla. An outgrowth from the epithelial column forms a *sebaceous gland*.

BIRTH

At birth a hair projects from the follicle, and the secretions of the sebaceous gland lubricate the hair shaft.

EXOCRINE GLANDS

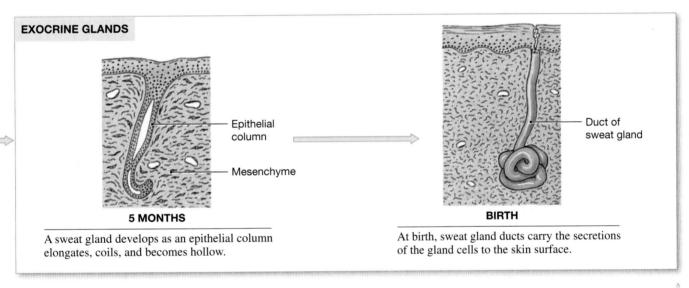

5 MONTHS

A sweat gland develops as an epithelial column elongates, coils, and becomes hollow.

BIRTH

At birth, sweat gland ducts carry the secretions of the gland cells to the skin surface.

MAMMARY GLANDS

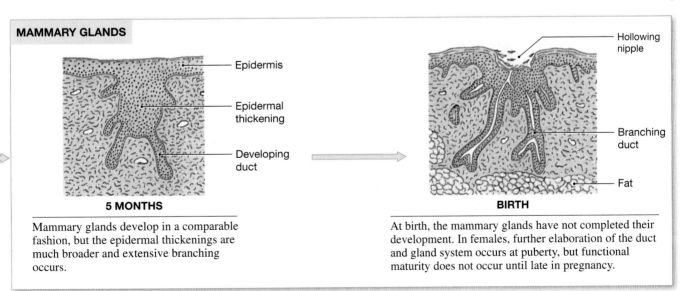

5 MONTHS

Mammary glands develop in a comparable fashion, but the epidermal thickenings are much broader and extensive branching occurs.

BIRTH

At birth, the mammary glands have not completed their development. In females, further elaboration of the duct and gland system occurs at puberty, but functional maturity does not occur until late in pregnancy.

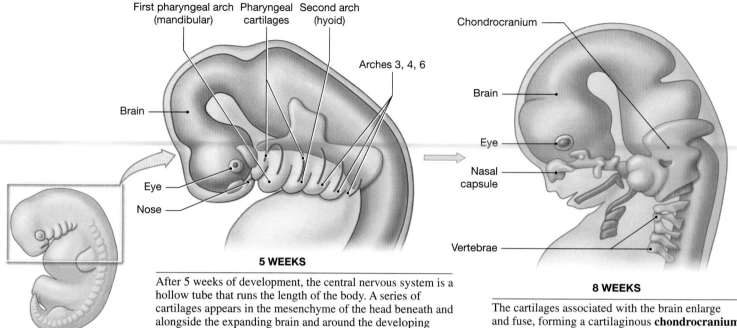

5 WEEKS

After 5 weeks of development, the central nervous system is a hollow tube that runs the length of the body. A series of cartilages appears in the mesenchyme of the head beneath and alongside the expanding brain and around the developing nose, eyes, and ears. These cartilages are shown in light blue. Five additional pairs of cartilages develop in the walls of the pharynx. These cartilages, shown in dark blue, are located within the **pharyngeal**, or **branchial, arches**. (*Branchial* refers to gills—in fish the caudal arches develop into skeletal supports for the gills.) The first arch, or **mandibular arch**, is the largest.

8 WEEKS

The cartilages associated with the brain enlarge and fuse, forming a cartilaginous **chondrocranium** (kon-drō-KRĀ-nē-um; *chondros*, cartilage + *cranium*, skull) that cradles the brain and sense organs. At 8 weeks its walls and floor are incomplete, and there is no roof.

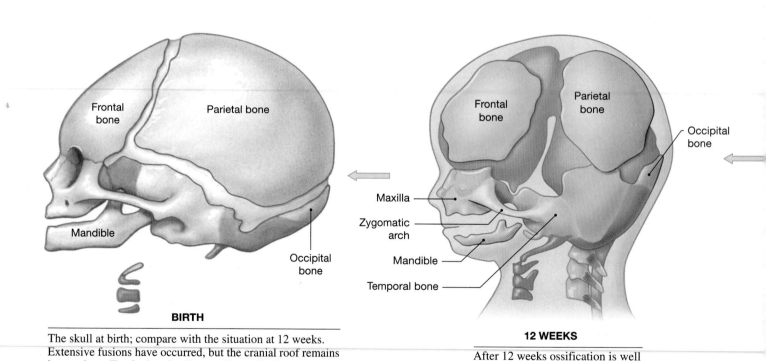

BIRTH

The skull at birth; compare with the situation at 12 weeks. Extensive fusions have occurred, but the cranial roof remains incomplete. (For further details, *see Figure 7–15, p. 230 of the text.*)

12 WEEKS

After 12 weeks ossification is well under way in the cranium and face. Compare with Plate 90b, page 114.

EMBRYOLOGY SUMMARY 6: THE DEVELOPMENT OF THE SKULL

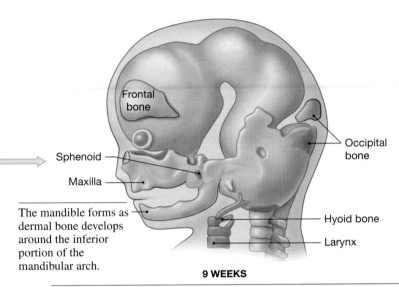

9 WEEKS

Sphenoid

Maxilla

The mandible forms as dermal bone develops around the inferior portion of the mandibular arch.

Frontal bone

Occipital bone

Hyoid bone

Larynx

During the ninth week, numerous centers of endochondral ossification appear within the chondrocranium. These centers are shown in red. Gradually, the frontal and parietal bones of the cranial roof appear as intramembranous ossification begins in the overlying dermis. As these centers (beige) enlarge and expand, extensive fusions occur.

The dorsal portion of the mandibular arch fuses with the chondrocranium. The fused cartilages do not ossify; instead, osteoblasts begin sheathing them in dermal bone. On each side this sheath fuses with a bone developing at the entrance to the nasal cavity, producing the two maxillae. Ossification centers in the roof of the mouth spread to form the palatine processes and later fuse with the maxillae.

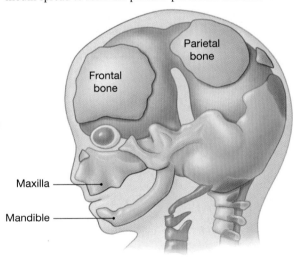

Parietal bone

Frontal bone

Maxilla

Mandible

10 WEEKS

The second arch, or **hyoid arch**, forms near the temporal bones. Fusion of the superior tips of the hyoid with the temporals forms the styloid processes. The ventral portion of the hyoid arch ossifies as the hyoid bone. The third arch fuses with the hyoid, and the fourth and sixth arches form laryngeal cartilages.

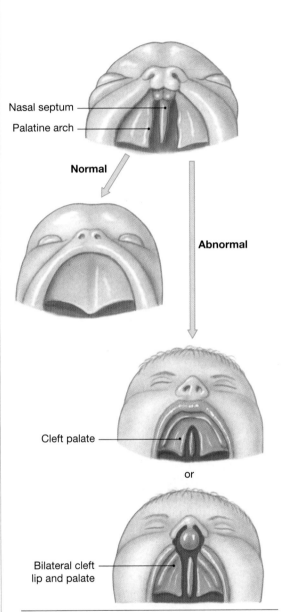

Nasal septum

Palatine arch

Normal

Abnormal

Cleft palate

or

Bilateral cleft lip and palate

If the overlying skin does not fuse normally, the result is a **cleft lip**. Cleft lips affect roughly one birth in a thousand. A split extending into the orbit and palate is called a **cleft palate**. Cleft palates are half as common as cleft lips. Both conditions can be corrected surgically.

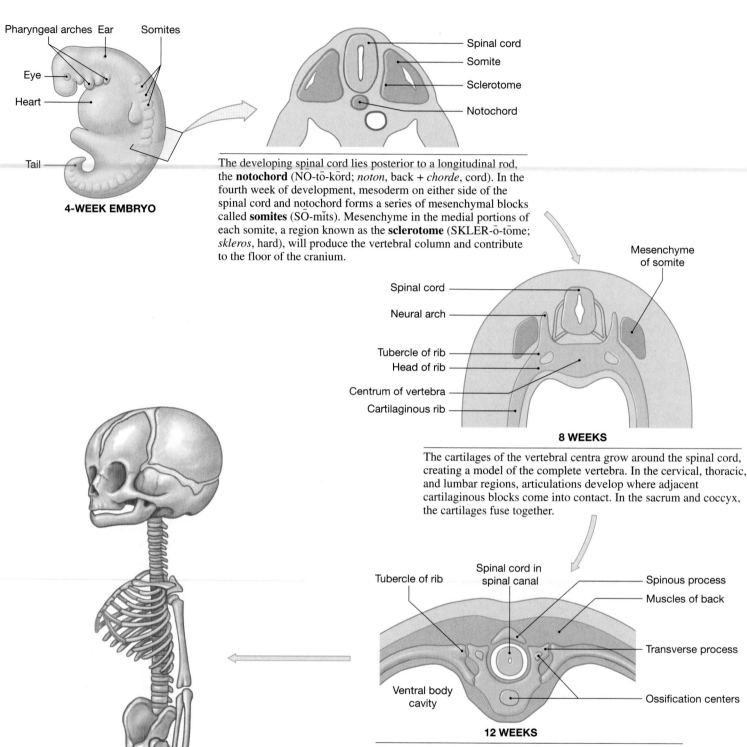

Pharyngeal arches Ear   Somites

Eye

Heart

Tail

4-WEEK EMBRYO

Spinal cord
Somite
Sclerotome
Notochord

The developing spinal cord lies posterior to a longitudinal rod, the **notochord** (NŌ-tō-kōrd; *noton*, back + *chorde*, cord). In the fourth week of development, mesoderm on either side of the spinal cord and notochord forms a series of mesenchymal blocks called **somites** (SŌ-mīts). Mesenchyme in the medial portions of each somite, a region known as the **sclerotome** (SKLER-ō-tōme; *skleros*, hard), will produce the vertebral column and contribute to the floor of the cranium.

Mesenchyme
of somite

Spinal cord

Neural arch

Tubercle of rib
Head of rib
Centrum of vertebra
Cartilaginous rib

8 WEEKS

The cartilages of the vertebral centra grow around the spinal cord, creating a model of the complete vertebra. In the cervical, thoracic, and lumbar regions, articulations develop where adjacent cartilaginous blocks come into contact. In the sacrum and coccyx, the cartilages fuse together.

Tubercle of rib

Spinal cord in
spinal canal

Spinous process
Muscles of back

Transverse process

Ventral body
cavity

Ossification centers

12 WEEKS

About the time the ribs separate from the vertebrae, ossification begins. Only the shortest ribs undergo complete ossification. In the rest, the distal portions remain cartilaginous, forming the costal cartilages. Several ossification centers appear in the sternum, but fusion gradually reduces the number.

BIRTH

At birth, the vertebrae and ribs are ossified, but many cartilaginous areas remain. For example, the anterior portions of the ribs remain cartilaginous. Additional growth will occur for many years; in vertebrae, the bases of the neural arches enlarge until ages 3–6, and the spinal processes and vertebral bodies grow until ages 18–25.

EMBRYOLOGY SUMMARY 7: THE DEVELOPMENT OF THE VERTEBRAL COLUMN

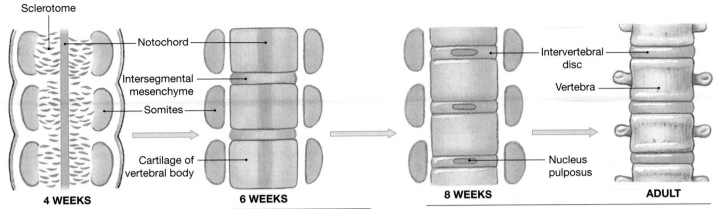

Sclerotome

Notochord

Intersegmental mesenchyme

Somites

Cartilage of vertebral body

4 WEEKS

6 WEEKS

Intervertebral disc

Vertebra

Nucleus pulposus

8 WEEKS

ADULT

Cells of the sclerotomal segments migrate away from the somites and cluster around the notochord.

The migrating cells differentiate into chondroblasts and produce a series of cartilaginous blocks that surround the notochord. These cartilages, which will develop into the vertebral centra, are separated by patches of mesenchyme.

Expansion of the vertebral centra eventually eliminates the notochord, but it remains intact between adjacent vertebrae, forming the *nucleus pulposus* of the intervertebral discs. Later, surrounding mesenchymal cells differentiate into chondroblasts and produce the fibrous cartilage of the *anulus fibrosus*.

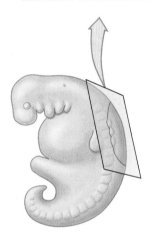

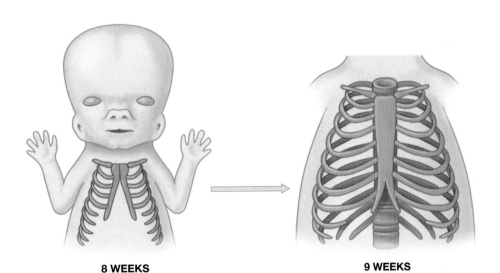

8 WEEKS

9 WEEKS

Rib cartilages expand away from the developing transverse processes of the vertebrae. At first they are continuous, but by week 8 the ribs have separated from the vertebrae. Ribs form at every vertebra, but in the cervical, lumbar, sacral, and coccygeal regions they remain small and later fuse with the growing vertebrae. The ribs of the thoracic vertebrae continue to enlarge, following the curvature of the body wall. When they reach the ventral midline, they fuse with the cartilages of the sternum.

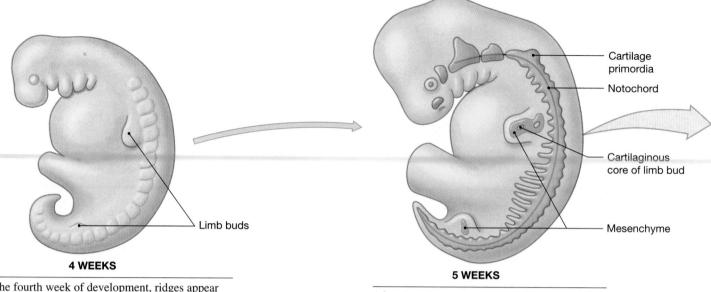

4 WEEKS

In the fourth week of development, ridges appear along the flanks of the embryo, extending from just behind the throat to just before the anus. These ridges form as mesodermal cells congregate beneath the ectoderm of the flank. Mesoderm gradually accumulates at the end of each ridge, forming two pairs of limb buds.

5 WEEKS

After 5 weeks of development, the pectoral limb buds have a cartilaginous core and scapular cartilages are developing in the mesenchyme of the trunk.

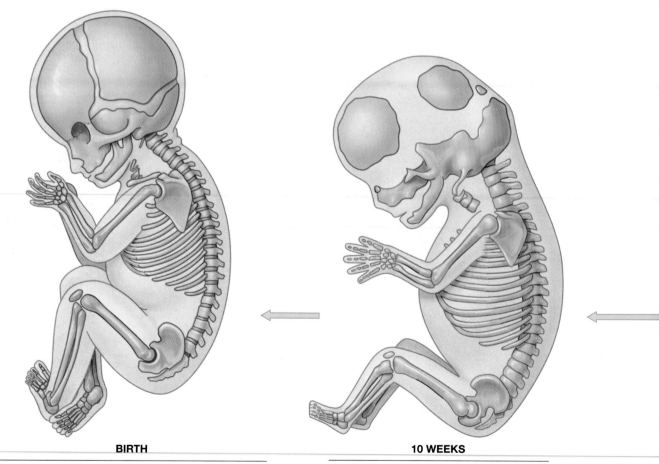

BIRTH

The skeleton of a newborn infant. Note the extensive areas of cartilage (blue) in the humeral head, in the wrist, between the bones of the palm and fingers, and in the hips. Notice the appearance of the axial skeleton, with reference to the two previous Embryology Summaries.

10 WEEKS

Ossification in the embryonic skeleton after approximately 10 weeks of development. The shafts of the limb bones are undergoing rapid ossification, but the distal bones of the carpus and tarsus remain cartilaginous.

EMBRYOLOGY SUMMARY 8: THE DEVELOPMENT OF THE APPENDICULAR SKELETON

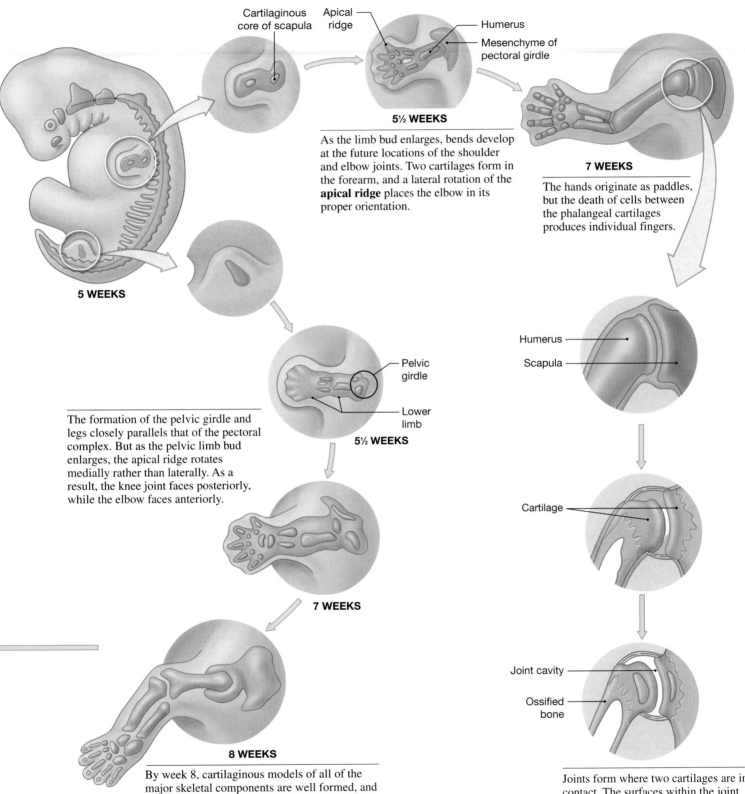

Cartilaginous core of scapula

Apical ridge

Humerus

Mesenchyme of pectoral girdle

5½ WEEKS

As the limb bud enlarges, bends develop at the future locations of the shoulder and elbow joints. Two cartilages form in the forearm, and a lateral rotation of the **apical ridge** places the elbow in its proper orientation.

7 WEEKS

The hands originate as paddles, but the death of cells between the phalangeal cartilages produces individual fingers.

5 WEEKS

Pelvic girdle

Lower limb

5½ WEEKS

The formation of the pelvic girdle and legs closely parallels that of the pectoral complex. But as the pelvic limb bud enlarges, the apical ridge rotates medially rather than laterally. As a result, the knee joint faces posteriorly, while the elbow faces anteriorly.

7 WEEKS

Humerus

Scapula

Cartilage

Joint cavity

Ossified bone

8 WEEKS

By week 8, cartilaginous models of all of the major skeletal components are well formed, and endochondral ossification begins in the future limb bones. Ossification of the hip bones begins at three separate centers that gradually enlarge.

Joints form where two cartilages are in contact. The surfaces within the joint cavity remain cartilaginous, while the rest of the bones undergo ossification.

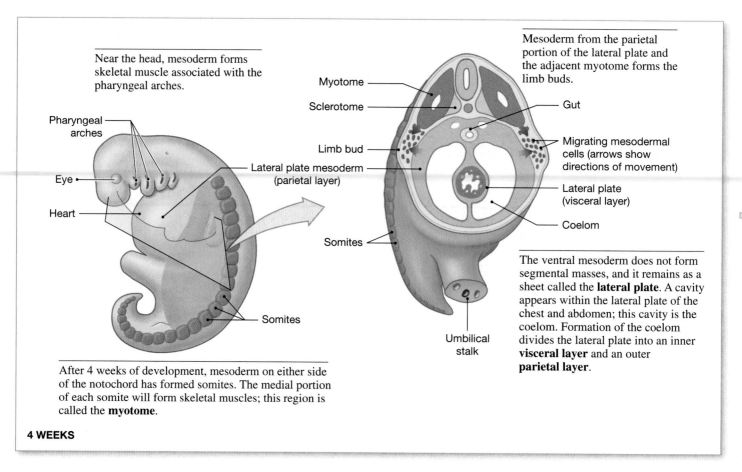

Near the head, mesoderm forms skeletal muscle associated with the pharyngeal arches.

Pharyngeal arches

Eye

Heart

Somites

Lateral plate mesoderm (parietal layer)

After 4 weeks of development, mesoderm on either side of the notochord has formed somites. The medial portion of each somite will form skeletal muscles; this region is called the **myotome**.

4 WEEKS

Mesoderm from the parietal portion of the lateral plate and the adjacent myotome forms the limb buds.

Myotome

Sclerotome

Limb bud

Somites

Gut

Migrating mesodermal cells (arrows show directions of movement)

Lateral plate (visceral layer)

Coelom

Umbilical stalk

The ventral mesoderm does not form segmental masses, and it remains as a sheet called the **lateral plate**. A cavity appears within the lateral plate of the chest and abdomen; this cavity is the coelom. Formation of the coelom divides the lateral plate into an inner **visceral layer** and an outer **parietal layer**.

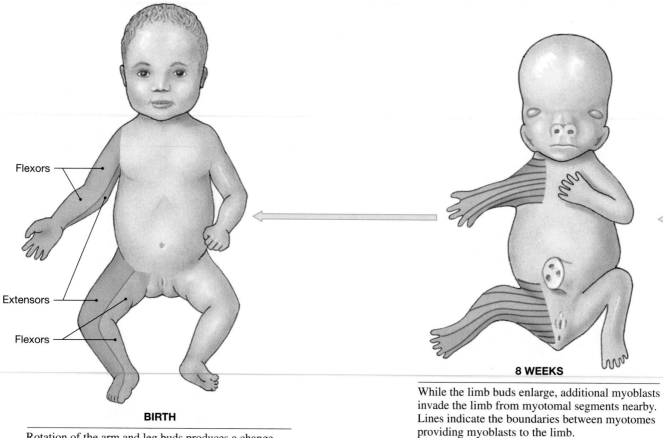

Flexors

Extensors

Flexors

BIRTH

Rotation of the arm and leg buds produces a change in the position of these masses relative to the body axis.

8 WEEKS

While the limb buds enlarge, additional myoblasts invade the limb from myotomal segments nearby. Lines indicate the boundaries between myotomes providing myoblasts to the limb.

EMBRYOLOGY SUMMARY 9: THE DEVELOPMENT OF THE MUSCULAR SYSTEM

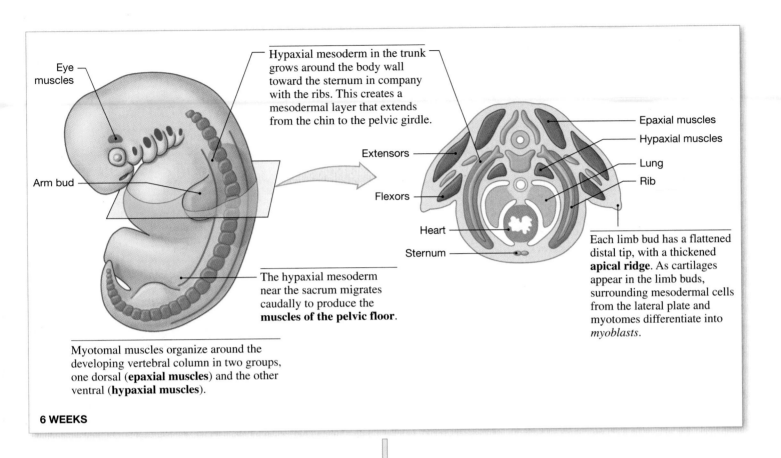

Eye muscles

Arm bud

Hypaxial mesoderm in the trunk grows around the body wall toward the sternum in company with the ribs. This creates a mesodermal layer that extends from the chin to the pelvic girdle.

Extensors

Flexors

Heart

Sternum

Epaxial muscles

Hypaxial muscles

Lung

Rib

Each limb bud has a flattened distal tip, with a thickened **apical ridge**. As cartilages appear in the limb buds, surrounding mesodermal cells from the lateral plate and myotomes differentiate into *myoblasts*.

The hypaxial mesoderm near the sacrum migrates caudally to produce the **muscles of the pelvic floor**.

Myotomal muscles organize around the developing vertebral column in two groups, one dorsal (**epaxial muscles**) and the other ventral (**hypaxial muscles**).

6 WEEKS

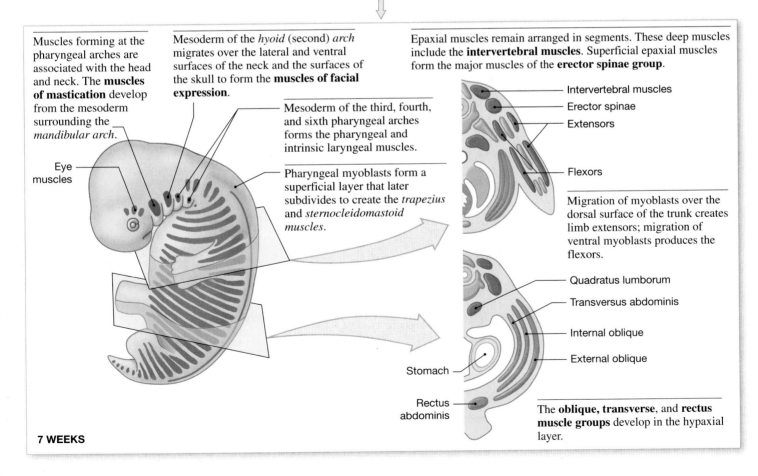

Muscles forming at the pharyngeal arches are associated with the head and neck. The **muscles of mastication** develop from the mesoderm surrounding the *mandibular arch*.

Eye muscles

Mesoderm of the *hyoid* (second) *arch* migrates over the lateral and ventral surfaces of the neck and the surfaces of the skull to form the **muscles of facial expression**.

Mesoderm of the third, fourth, and sixth pharyngeal arches forms the pharyngeal and intrinsic laryngeal muscles.

Pharyngeal myoblasts form a superficial layer that later subdivides to create the *trapezius* and *sternocleidomastoid muscles*.

Epaxial muscles remain arranged in segments. These deep muscles include the **intervertebral muscles**. Superficial epaxial muscles form the major muscles of the **erector spinae group**.

Intervertebral muscles

Erector spinae

Extensors

Flexors

Migration of myoblasts over the dorsal surface of the trunk creates limb extensors; migration of ventral myoblasts produces the flexors.

Quadratus lumborum

Transversus abdominis

Internal oblique

External oblique

Stomach

Rectus abdominis

The **oblique, transverse**, and **rectus muscle groups** develop in the hypaxial layer.

7 WEEKS

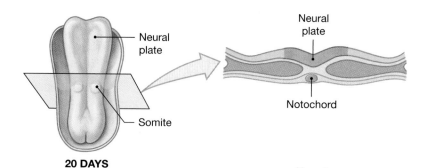

After two weeks of development, *somites* are appearing on either side of the *notochord*. The ectoderm near the midline thickens, forming an elevated neural plate. The **neural plate** is largest near the future head of the developing embryo.

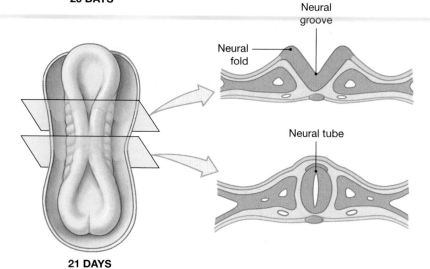

A crease develops along the axis of the neural plate, creating the **neural groove**. The edges, or **neural folds**, gradually move together. They first contact one another midway along the axis of the neural plate, near the end of the third week.

Where the neural folds meet, they fuse to form a cylindrical **neural tube** that loses its connection with the superficial ectoderm. The process of neural tube formation is called **neurulation**; it is completed in less than a week. The formation of the axial skeleton and that of the musculature around the developing neural tube were described on pages 124–125 and 128–129.

Cells at the tips of the neural folds do not participate in neural tube formation. These cells of the **neural crest** at first remain between the dorsal surface of the neural tube and the ectoderm, but they later migrate to other locations. The neural tube becomes the CNS. Axons from neurons within the neural tube and the axons of neural crest cells form the PNS.

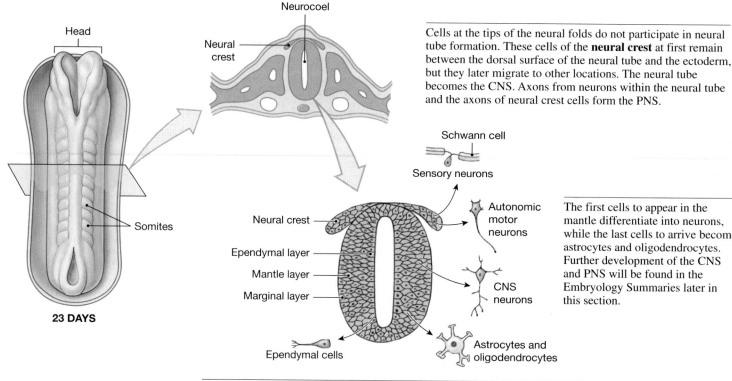

The first cells to appear in the mantle differentiate into neurons, while the last cells to arrive become astrocytes and oligodendrocytes. Further development of the CNS and PNS will be found in the Embryology Summaries later in this section.

The neural tube increases in thickness as its epithelial lining undergoes repeated mitoses. By the middle of the fifth developmental week, there are three distinct layers. The **ependymal layer** lines the enclosed cavity, or **neurocoel**. The ependymal cells continue their mitotic activities, and daughter cells create the surrounding **mantle layer**. Axons from developing neurons form a superficial **marginal layer**.

EMBRYOLOGY SUMMARY 10:   **AN INTRODUCTION TO THE DEVELOPMENT OF THE NERVOUS SYSTEM**

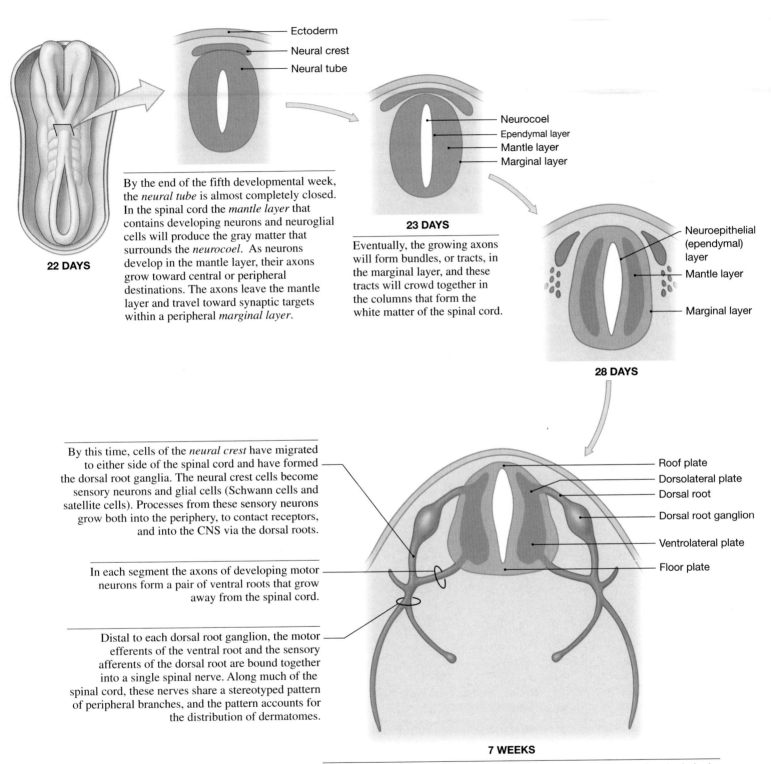

Ectoderm
Neural crest
Neural tube

22 DAYS

By the end of the fifth developmental week, the *neural tube* is almost completely closed. In the spinal cord the *mantle layer* that contains developing neurons and neuroglial cells will produce the gray matter that surrounds the *neurocoel*. As neurons develop in the mantle layer, their axons grow toward central or peripheral destinations. The axons leave the mantle layer and travel toward synaptic targets within a peripheral *marginal layer*.

23 DAYS

Neurocoel
Ependymal layer
Mantle layer
Marginal layer

Eventually, the growing axons will form bundles, or tracts, in the marginal layer, and these tracts will crowd together in the columns that form the white matter of the spinal cord.

28 DAYS

Neuroepithelial (ependymal) layer
Mantle layer
Marginal layer

By this time, cells of the *neural crest* have migrated to either side of the spinal cord and have formed the dorsal root ganglia. The neural crest cells become sensory neurons and glial cells (Schwann cells and satellite cells). Processes from these sensory neurons grow both into the periphery, to contact receptors, and into the CNS via the dorsal roots.

In each segment the axons of developing motor neurons form a pair of ventral roots that grow away from the spinal cord.

Distal to each dorsal root ganglion, the motor efferents of the ventral root and the sensory afferents of the dorsal root are bound together into a single spinal nerve. Along much of the spinal cord, these nerves share a stereotyped pattern of peripheral branches, and the pattern accounts for the distribution of dermatomes.

Roof plate
Dorsolateral plate
Dorsal root
Dorsal root ganglion
Ventrolateral plate
Floor plate

7 WEEKS

As the mantle enlarges, the neurocoel becomes laterally compressed and relatively narrow. The relatively thin **roof plate** and **floor plate** will not thicken substantially, but the **dorsolateral** and **ventrolateral plates** enlarge rapidly. Neurons developing within the dorsolateral plate will receive and relay sensory information, while those in the ventrolateral region will develop into motor neurons.

EMBRYOLOGY SUMMARY 11: THE DEVELOPMENT OF THE SPINAL CORD AND SPINAL NERVES—*PART I*

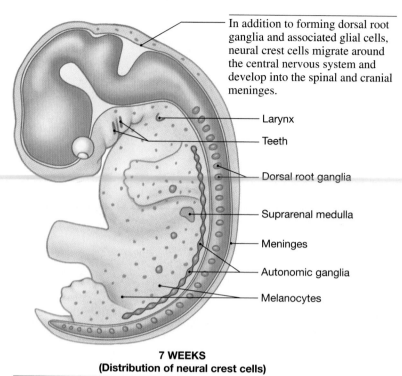

In addition to forming dorsal root ganglia and associated glial cells, neural crest cells migrate around the central nervous system and develop into the spinal and cranial meninges.

- Larynx
- Teeth
- Dorsal root ganglia
- Suprarenal medulla
- Meninges
- Autonomic ganglia
- Melanocytes

7 WEEKS
(Distribution of neural crest cells)

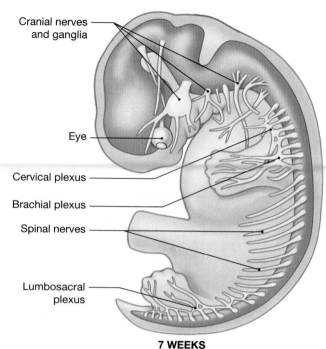

- Cranial nerves and ganglia
- Eye
- Cervical plexus
- Brachial plexus
- Spinal nerves
- Lumbosacral plexus

7 WEEKS
(Peripheral nerve distribution)

Neural crest cells aggregate to form autonomic ganglia near the vertebral column and in peripheral organs. Migrating neural crest cells contribute to the formation of teeth and form the laryngeal cartilages, melanocytes of the skin, the skull, connective tissues around the eye, the intrinsic muscles of the eye, Schwann cells, satellite cells, and the suprarenal medullae.

Several spinal nerves innervate each developing limb. When embryonic muscle cells migrate away from the myotome, the nerves grow right along with them. If a large muscle in the adult is derived from several myotomal blocks, connective tissue partitions will often mark the original boundaries, and the innervation will always involve more than one spinal nerve.

DEVELOPMENTAL ABNORMALITIES

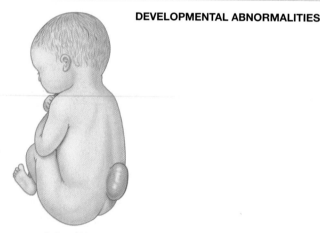

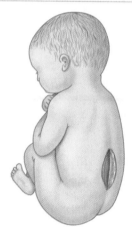

Spina bifida

Neural tube defect

Spina bifida (BI-fi-da) results when the developing vertebral laminae fail to unite due to abnormal neural tube formation at that site. The neural arch is incomplete, and the meninges bulge outward beneath the skin of the back. The extent of the abnormality determines the severity of the defects. In mild cases, the condition may pass unnoticed; extreme cases involve much of the length of the vertebral column.

A **neural tube defect (NTD)** is a condition that is secondary to a developmental error in the formation of the spinal cord. Instead of forming a hollow tube, a portion of the spinal cord develops as a broad plate. This is often associated with spina bifida. Neural tube defects affect roughly one individual in 1000; prenatal testing can detect the existence of these defects with an 80–85 percent success rate.

Before proceeding, briefly review the summaries of skull formation, vertebral column development, and development of the spinal cord in the previous Embryology Summaries.

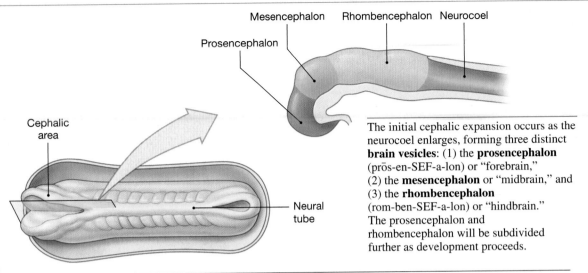

The initial cephalic expansion occurs as the neurocoel enlarges, forming three distinct **brain vesicles**: (1) the **prosencephalon** (prōs-en-SEF-a-lon) or "forebrain," (2) the **mesencephalon** or "midbrain," and (3) the **rhombencephalon** (rom-ben-SEF-a-lon) or "hindbrain." The prosencephalon and rhombencephalon will be subdivided further as development proceeds.

Even before **neural tube** formation has been completed, the cephalic portion begins to enlarge. Major differences in brain versus spinal cord development include (1) early breakdown of mantle (gray matter) and marginal (white matter) organization; (2) appearance of areas of neural cortex; (3) differential growth between and within specific regions; (4) appearance of characteristic bends and folds; and (5) loss of obvious segmental organization.

23 DAYS

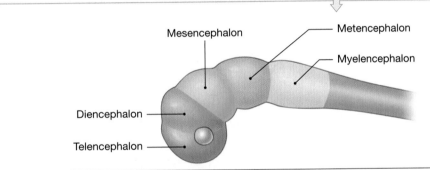

The rhombencephalon first subdivides into the **metencephalon** (met-en-SEF-a-lon; *meta*, after) and the **myelencephalon** (mi-el-ēn-SEF-a-lon; *myelon*, spinal cord).

The prosencephalon forms the **telecephalon** (tel-en-SEF-a-lon; *telos*, end + *enkephalos*, brain) and the **diencephalon**. The telencephalon begins as a pair of swellings near the rostral, dorsolateral border of the prosencephalon.

4 WEEKS

Cranial nerves develop as sensory ganglia and link peripheral receptors with the brain, and motor fibers grow out of developing cranial nuclei. Special sensory neurons of cranial nerves N I, II, and VIII develop in association with the developing receptors. The somatic motor nerves (N III, IV, and VI) grow to the eye muscles; the mixed nerves (N V, VII, IX, and X) innervate the **pharyngeal arches**.

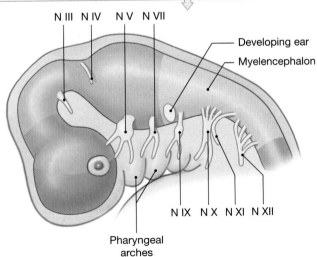

Development of the **mesencephalon** produces a small mass of neural tissue with a constricted neurocoel, the *aqueduct of the midbrain*.

As differential growth proceeds and the position and orientation of the embryo change, a series of bends, or **flexures** (FLEK-sherz), appears along the axis of the developing brain.

5 WEEKS

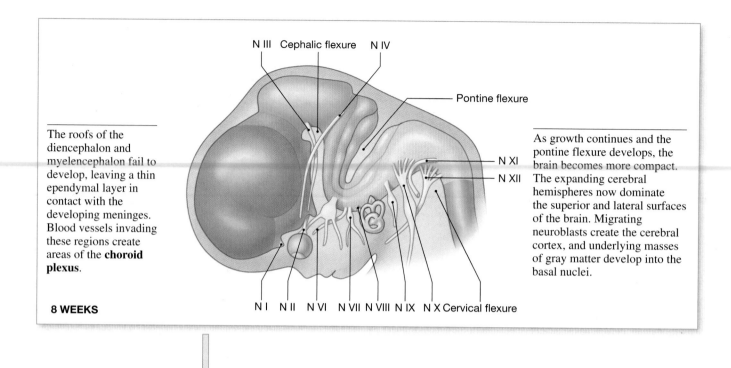

The roofs of the diencephalon and myelencephalon fail to develop, leaving a thin ependymal layer in contact with the developing meninges. Blood vessels invading these regions create areas of the **choroid plexus**.

8 WEEKS

As growth continues and the pontine flexure develops, the brain becomes more compact. The expanding cerebral hemispheres now dominate the superior and lateral surfaces of the brain. Migrating neuroblasts create the cerebral cortex, and underlying masses of gray matter develop into the basal nuclei.

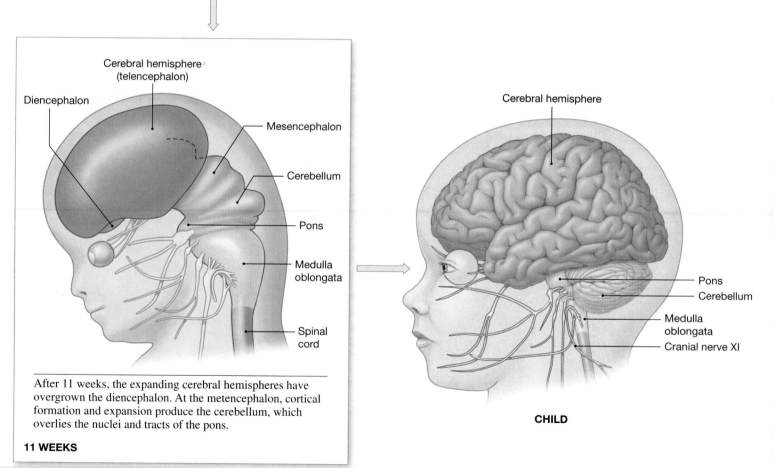

After 11 weeks, the expanding cerebral hemispheres have overgrown the diencephalon. At the metencephalon, cortical formation and expansion produce the cerebellum, which overlies the nuclei and tracts of the pons.

11 WEEKS

CHILD

EMBRYOLOGY SUMMARY **12**: THE DEVELOPMENT OF THE BRAIN AND CRANIAL NERVES—PART II

All special sense organs develop from the interaction between the epithelia and the developing nervous system of the embryo.

VISION

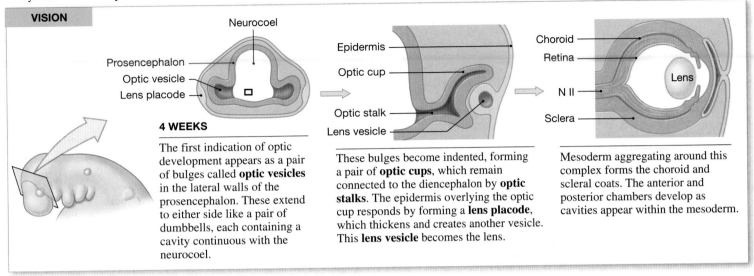

4 WEEKS

The first indication of optic development appears as a pair of bulges called **optic vesicles** in the lateral walls of the prosencephalon. These extend to either side like a pair of dumbbells, each containing a cavity continuous with the neurocoel.

These bulges become indented, forming a pair of **optic cups**, which remain connected to the diencephalon by **optic stalks**. The epidermis overlying the optic cup responds by forming a **lens placode**, which thickens and creates another vesicle. This **lens vesicle** becomes the lens.

Mesoderm aggregating around this complex forms the choroid and scleral coats. The anterior and posterior chambers develop as cavities appear within the mesoderm.

OLFACTION

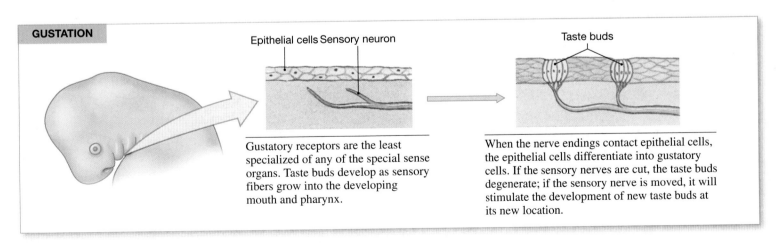

5 WEEKS

Olfactory receptors begin as a pair of thickened areas in front of the prosencephalon during the fifth developmental week. The thickenings are called **nasal placodes**.

10 WEEKS

Over time, the nasal placodes are enfolded and protected by developing facial structures. (Development of the face was discussed in the previous Embryology Summary of the Skull.)

GUSTATION

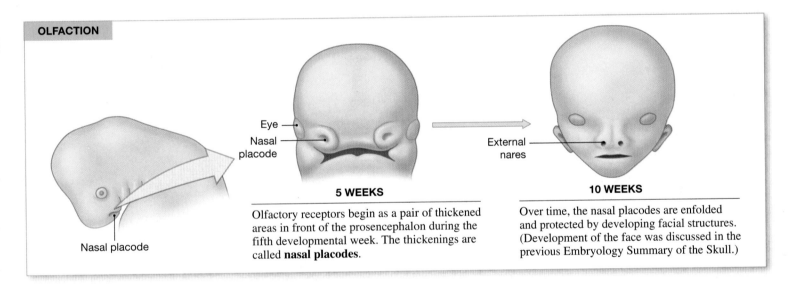

Gustatory receptors are the least specialized of any of the special sense organs. Taste buds develop as sensory fibers grow into the developing mouth and pharynx.

When the nerve endings contact epithelial cells, the epithelial cells differentiate into gustatory cells. If the sensory nerves are cut, the taste buds degenerate; if the sensory nerve is moved, it will stimulate the development of new taste buds at its new location.

EMBRYOLOGY SUMMARY 13: THE DEVELOPMENT OF SPECIAL SENSE ORGANS—*PART I*

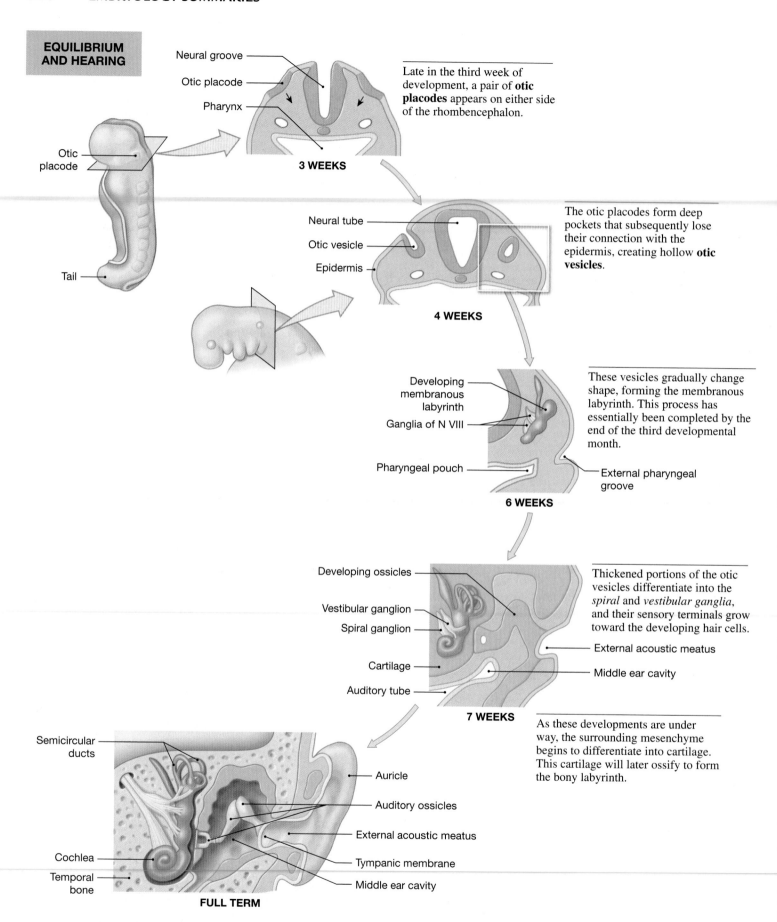

EQUILIBRIUM AND HEARING

Neural groove
Otic placode
Pharynx

3 WEEKS

Late in the third week of development, a pair of **otic placodes** appears on either side of the rhombencephalon.

Otic placode

Tail

Neural tube
Otic vesicle
Epidermis

4 WEEKS

The otic placodes form deep pockets that subsequently lose their connection with the epidermis, creating hollow **otic vesicles**.

Developing membranous labyrinth
Ganglia of N VIII

Pharyngeal pouch

External pharyngeal groove

6 WEEKS

These vesicles gradually change shape, forming the membranous labyrinth. This process has essentially been completed by the end of the third developmental month.

Developing ossicles

Vestibular ganglion
Spiral ganglion

Cartilage

Auditory tube

External acoustic meatus

Middle ear cavity

7 WEEKS

Thickened portions of the otic vesicles differentiate into the *spiral* and *vestibular ganglia*, and their sensory terminals grow toward the developing hair cells.

Semicircular ducts

Cochlea

Temporal bone

Auricle

Auditory ossicles

External acoustic meatus

Tympanic membrane

Middle ear cavity

FULL TERM

As these developments are under way, the surrounding mesenchyme begins to differentiate into cartilage. This cartilage will later ossify to form the bony labyrinth.

EMBRYOLOGY SUMMARY 13: THE DEVELOPMENT OF SPECIAL SENSE ORGANS—PART II

As noted in Chapter 4, all secretory glands, whether exocrine or endocrine, are derived from epithelia. Endocrine organs develop from epithelia (1) covering the outside of the embryo, (2) lining the digestive tract, and (3) lining the coelomic cavity.

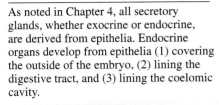

Pharyngeal arches **Pharyngeal clefts**

WEEK 5

The pharyngeal region of the embryo plays a particularly important role in endocrine development. After 4–5 weeks of development, the *pharyngeal arches* are well formed. Human embryos develop five or six pharyngeal arches, not all visible from the exterior. (Arch 5 may not appear or may form and degenerate almost immediately.) The five major arches (I–IV, VI) are separated by *pharyngeal clefts*, deep ectodermal grooves.

PARATHYROID GLANDS AND THYMUS

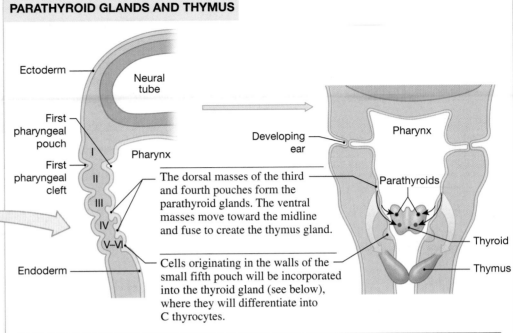

Ectoderm

Neural tube

First pharyngeal pouch

First pharyngeal cleft

Pharynx

I
II
III
IV
V–VI

Endoderm

Developing ear

Pharynx

Parathyroids

Thyroid

Thymus

The dorsal masses of the third and fourth pouches form the parathyroid glands. The ventral masses move toward the midline and fuse to create the thymus gland.

Cells originating in the walls of the small fifth pouch will be incorporated into the thyroid gland (see below), where they will differentiate into C thyrocytes.

In sectional view, five **pharyngeal pouches** extend laterally toward the pharyngeal clefts. The first pouch lies caudal to the first (mandibular) arch. Pharyngeal pouches 5 and 6 are very small and are interconnected. Endoderm lining the third, fourth, and fifth pairs of pharyngeal pouches forms dorsal and ventral masses of cells that migrate beneath the endodermal epithelium.

THYROID GLAND

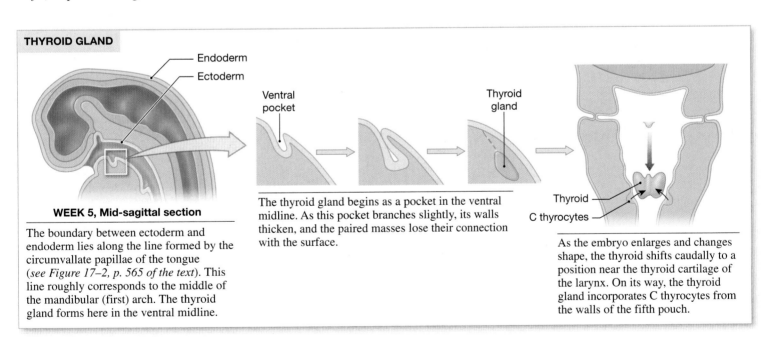

Endoderm
Ectoderm

Ventral pocket

Thyroid gland

Thyroid
C thyrocytes

WEEK 5, Mid-sagittal section

The boundary between ectoderm and endoderm lies along the line formed by the circumvallate papillae of the tongue (*see Figure 17–2, p. 565 of the text*). This line roughly corresponds to the middle of the mandibular (first) arch. The thyroid gland forms here in the ventral midline.

The thyroid gland begins as a pocket in the ventral midline. As this pocket branches slightly, its walls thicken, and the paired masses lose their connection with the surface.

As the embryo enlarges and changes shape, the thyroid shifts caudally to a position near the thyroid cartilage of the larynx. On its way, the thyroid gland incorporates C thyrocytes from the walls of the fifth pouch.

PITUITARY GLAND

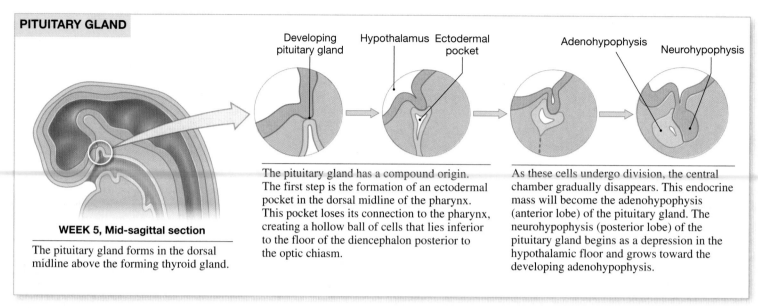

WEEK 5, Mid-sagittal section

The pituitary gland forms in the dorsal midline above the forming thyroid gland.

The pituitary gland has a compound origin. The first step is the formation of an ectodermal pocket in the dorsal midline of the pharynx. This pocket loses its connection to the pharynx, creating a hollow ball of cells that lies inferior to the floor of the diencephalon posterior to the optic chiasm.

As these cells undergo division, the central chamber gradually disappears. This endocrine mass will become the adenohypophysis (anterior lobe) of the pituitary gland. The neurohypophysis (posterior lobe) of the pituitary gland begins as a depression in the hypothalamic floor and grows toward the developing adenohypophysis.

SUPRARENAL GLANDS

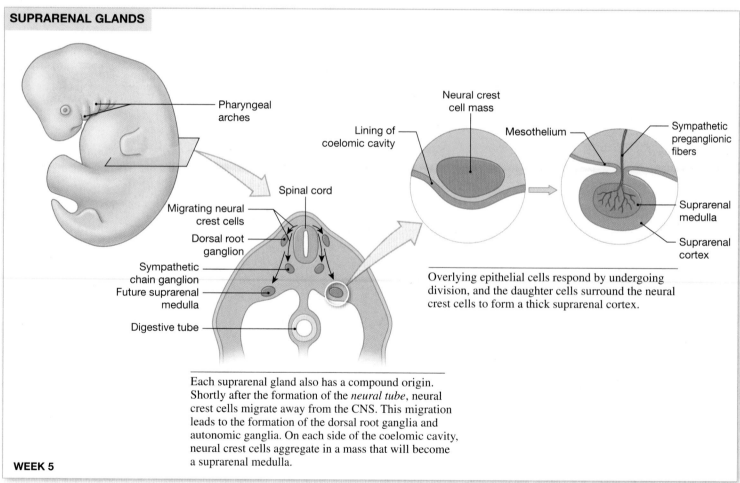

Overlying epithelial cells respond by undergoing division, and the daughter cells surround the neural crest cells to form a thick suprarenal cortex.

Each suprarenal gland also has a compound origin. Shortly after the formation of the *neural tube*, neural crest cells migrate away from the CNS. This migration leads to the formation of the dorsal root ganglia and autonomic ganglia. On each side of the coelomic cavity, neural crest cells aggregate in a mass that will become a suprarenal medulla.

WEEK 5

For additional details concerning the development of other endocrine organs, refer to the subsequent Embryology Summaries on the Lymphoid, Digestive, Urinary, and Reproductive systems.

EMBRYOLOGY SUMMARY 14: THE DEVELOPMENT OF THE ENDOCRINE SYSTEM—*PART II*

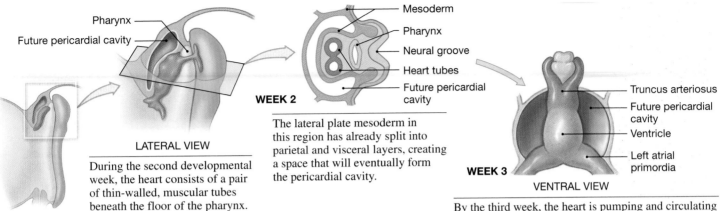

Pharynx
Future pericardial cavity

LATERAL VIEW

During the second developmental week, the heart consists of a pair of thin-walled, muscular tubes beneath the floor of the pharynx.

WEEK 2

Mesoderm
Pharynx
Neural groove
Heart tubes
Future pericardial cavity

The lateral plate mesoderm in this region has already split into parietal and visceral layers, creating a space that will eventually form the pericardial cavity.

WEEK 3

VENTRAL VIEW

Truncus arteriosus
Future pericardial cavity
Ventricle
Left atrial primordia

By the third week, the heart is pumping and circulating blood. The cardiac tubes have fused, producing a heart with a single central chamber. Two large veins bring blood to the heart, and a single large artery, the **truncus arteriosus**, carries blood to the general circulation.

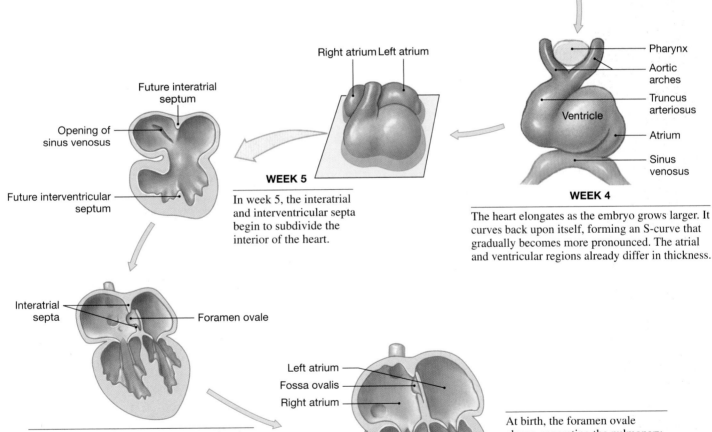

Future interatrial septum

Opening of sinus venosus

Future interventricular septum

Right atrium **Left atrium**

WEEK 5

In week 5, the interatrial and interventricular septa begin to subdivide the interior of the heart.

Pharynx
Aortic arches
Truncus arteriosus
Ventricle
Atrium
Sinus venosus

WEEK 4

The heart elongates as the embryo grows larger. It curves back upon itself, forming an S-curve that gradually becomes more pronounced. The atrial and ventricular regions already differ in thickness.

Interatrial septa
Foramen ovale

Two interatrial septa develop, one overlapping the other. A gap between the two, called the **foramen ovale**, permits blood flow from the right atrium to the left atrium. Backflow from left to right is prevented by a flap that acts as a one-way valve. Until birth, this atrial short circuit diverts blood from the pulmonary circuit.

Left atrium
Fossa ovalis
Right atrium

Right ventricle
Left ventricle

AGE 1 YEAR

At birth, the foramen ovale closes, separating the pulmonary and systemic circuits in the heart. A shallow depression, the **fossa ovalis**, remains through adulthood at the site of the foramen ovale. (*Other cardiovascular changes at birth are detailed in Figure 21–33, p. 768 of the text.*)

EMBRYOLOGY SUMMARY 15: THE DEVELOPMENT OF THE HEART

THE AORTIC ARCHES

An **aortic arch** carries arterial blood through each of the *pharyngeal arches*. In the dorsal pharyngeal wall, these vessels fuse to create the **dorsal aorta**, which distributes blood throughout the body. The arches are usually numbered from I to VI, corresponding to the pharyngeal arches.

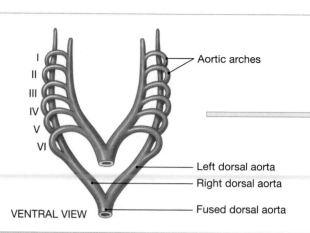

I
II
III
IV
V
VI

Aortic arches

Left dorsal aorta
Right dorsal aorta
Fused dorsal aorta

VENTRAL VIEW

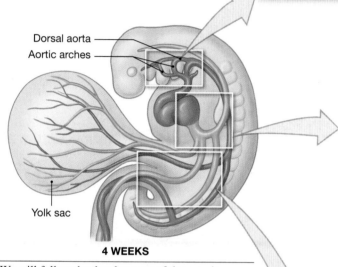

Dorsal aorta
Aortic arches

Yolk sac

4 WEEKS

We will follow the development of three major vessel complexes: the aortic arch, the venae cavae, and the hepatic portal and umbilical systems. (Arteries are shown in red and veins in blue regardless of the oxygenation of the blood they carry.)

THE VENAE CAVAE

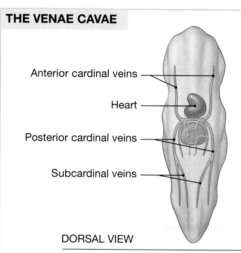

Anterior cardinal veins

Heart

Posterior cardinal veins

Subcardinal veins

DORSAL VIEW

The early venous circulation draining the tissues of the body wall, limbs, and head centers around the paired **anterior cardinal veins, posterior cardinal veins**, and **subcardinal veins**.

THE HEPATIC PORTAL AND UMBILICAL VESSELS

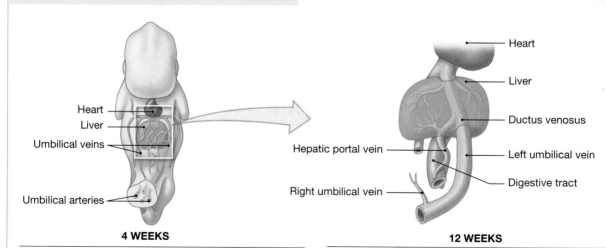

Heart
Liver
Umbilical veins

Umbilical arteries

4 WEEKS

Heart

Liver

Ductus venosus

Hepatic portal vein

Left umbilical vein

Right umbilical vein

Digestive tract

12 WEEKS

Paired **umbilical arteries** deliver blood to the placenta. At 4 weeks, paired **umbilical veins** return blood to capillary networks in the liver. Veins running along the length of the digestive tract have extensive interconnections.

By week 12, the right umbilical vein disintegrates, and the blood from the placenta travels along a single umbilical vein. The **ductus venosus** allows some venous blood to bypass the liver. The veins draining the digestive tract have fused, forming the hepatic portal vein.

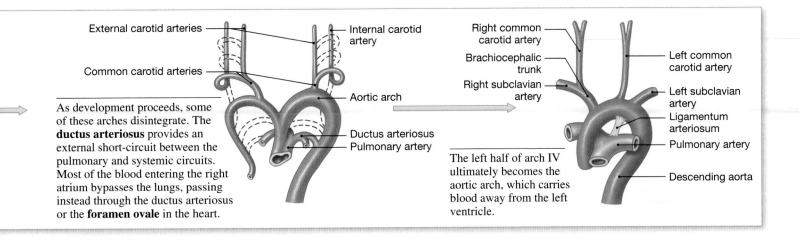

As development proceeds, some of these arches disintegrate. The **ductus arteriosus** provides an external short-circuit between the pulmonary and systemic circuits. Most of the blood entering the right atrium bypasses the lungs, passing instead through the ductus arteriosus or the **foramen ovale** in the heart.

The left half of arch IV ultimately becomes the aortic arch, which carries blood away from the left ventricle.

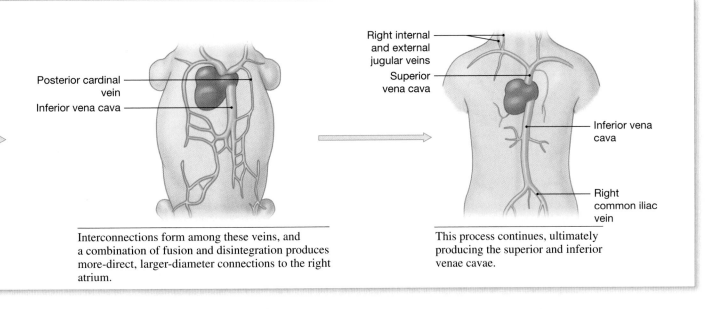

Interconnections form among these veins, and a combination of fusion and disintegration produces more-direct, larger-diameter connections to the right atrium.

This process continues, ultimately producing the superior and inferior venae cavae.

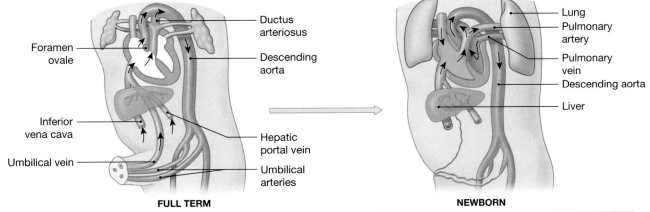

Shortly before birth, blood returning from the placenta travels through the liver in the ductus venosus to reach the inferior vena cava. Much of the blood delivered by the venae cavae bypasses the lungs by traveling through the foramen ovale and the ductus arteriosus.

At birth, pressures drop in the pleural cavities as the chest expands and the infant takes its first breath. The pulmonary vessels dilate, and blood flow to the lungs increases. Pressure falls in the right atrium, and the higher left atrial pressures close the valve that guards the foramen ovale. Smooth muscles contract the ductus arteriosus, which ultimately converts to the **ligamentum arteriosum**, a fibrous strand.

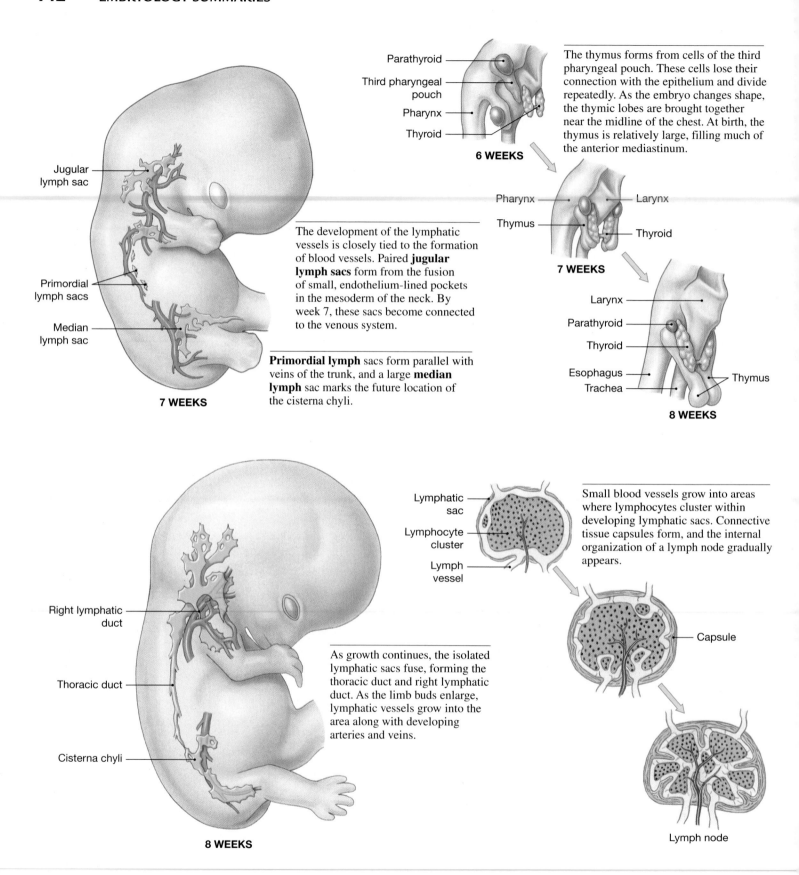

Parathyroid

Third pharyngeal pouch

Pharynx

Thyroid

6 WEEKS

The thymus forms from cells of the third pharyngeal pouch. These cells lose their connection with the epithelium and divide repeatedly. As the embryo changes shape, the thymic lobes are brought together near the midline of the chest. At birth, the thymus is relatively large, filling much of the anterior mediastinum.

Pharynx

Thymus

Larynx

Thyroid

7 WEEKS

Larynx

Parathyroid

Thyroid

Esophagus

Trachea

Thymus

8 WEEKS

Jugular lymph sac

Primordial lymph sacs

Median lymph sac

The development of the lymphatic vessels is closely tied to the formation of blood vessels. Paired **jugular lymph sacs** form from the fusion of small, endothelium-lined pockets in the mesoderm of the neck. By week 7, these sacs become connected to the venous system.

Primordial lymph sacs form parallel with veins of the trunk, and a large **median lymph** sac marks the future location of the cisterna chyli.

7 WEEKS

Right lymphatic duct

Thoracic duct

Cisterna chyli

8 WEEKS

As growth continues, the isolated lymphatic sacs fuse, forming the thoracic duct and right lymphatic duct. As the limb buds enlarge, lymphatic vessels grow into the area along with developing arteries and veins.

Lymphatic sac

Lymphocyte cluster

Lymph vessel

Small blood vessels grow into areas where lymphocytes cluster within developing lymphatic sacs. Connective tissue capsules form, and the internal organization of a lymph node gradually appears.

Capsule

Lymph node

EMBRYOLOGY SUMMARY 17: THE DEVELOPMENT OF THE LYMPHATIC SYSTEM

THE LUNGS

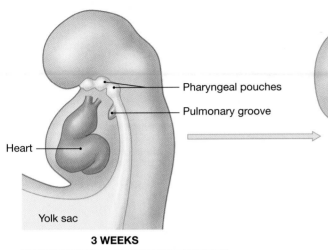

3 WEEKS

A shallow **pulmonary groove** appears in the midventral floor of the pharynx after roughly 3½ weeks of development. This groove, which lies near the level of the last pharyngeal arch, gradually deepens.

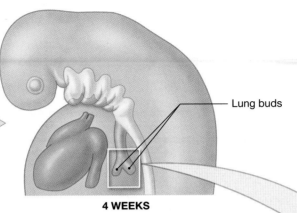

4 WEEKS

By week 4, the groove has become a blind pocket that extends caudally, anterior to the esophagus. This tube will become the trachea. At its tip, the tube branches, forming a pair of **lung buds**.

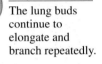

The lung buds continue to elongate and branch repeatedly.

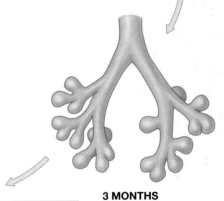

3 MONTHS

By the end of the sixth fetal month, there are around a million terminal branches, and the conducting passageways are complete to the level of the bronchioles.

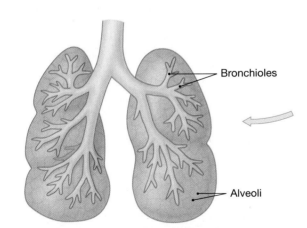

Over the next three months, each of the bronchioles gives rise to several hundred alveoli. This process continues for a variable period after birth.

EMBRYOLOGY SUMMARY 18: THE DEVELOPMENT OF THE RESPIRATORY SYSTEM—*PART I*

THE PLEURAL CAVITIES

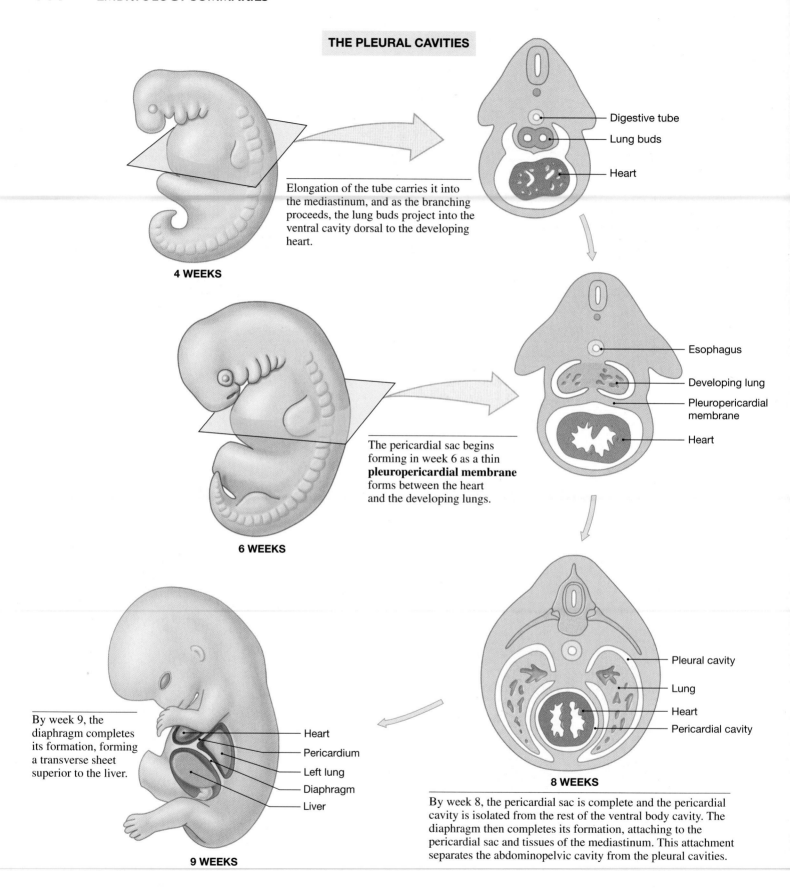

Elongation of the tube carries it into the mediastinum, and as the branching proceeds, the lung buds project into the ventral cavity dorsal to the developing heart.

4 WEEKS

Digestive tube

Lung buds

Heart

The pericardial sac begins forming in week 6 as a thin **pleuropericardial membrane** forms between the heart and the developing lungs.

6 WEEKS

Esophagus

Developing lung

Pleuropericardial membrane

Heart

By week 9, the diaphragm completes its formation, forming a transverse sheet superior to the liver.

Heart

Pericardium

Left lung

Diaphragm

Liver

9 WEEKS

Pleural cavity

Lung

Heart

Pericardial cavity

8 WEEKS

By week 8, the pericardial sac is complete and the pericardial cavity is isolated from the rest of the ventral body cavity. The diaphragm then completes its formation, attaching to the pericardial sac and tissues of the mediastinum. This attachment separates the abdominopelvic cavity from the pleural cavities.

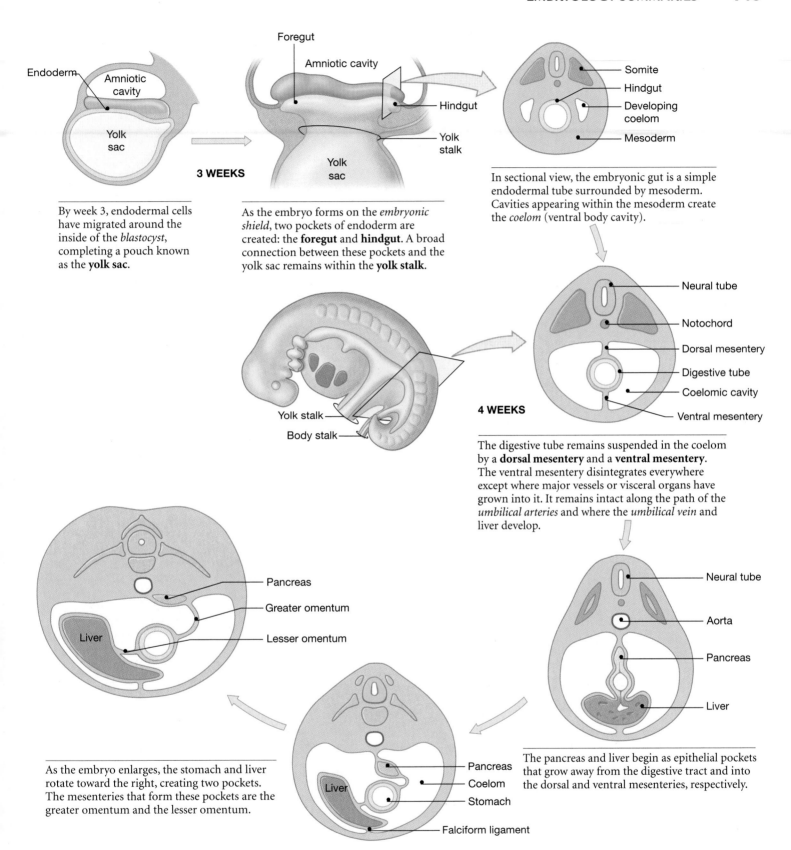

Endoderm

Amniotic cavity

Yolk sac

3 WEEKS

By week 3, endodermal cells have migrated around the inside of the *blastocyst*, completing a pouch known as the **yolk sac**.

Foregut

Amniotic cavity

Hindgut

Yolk stalk

Yolk sac

As the embryo forms on the *embryonic shield*, two pockets of endoderm are created: the **foregut** and **hindgut**. A broad connection between these pockets and the yolk sac remains within the **yolk stalk**.

Somite

Hindgut

Developing coelom

Mesoderm

In sectional view, the embryonic gut is a simple endodermal tube surrounded by mesoderm. Cavities appearing within the mesoderm create the *coelom* (ventral body cavity).

Yolk stalk

Body stalk

Neural tube

Notochord

Dorsal mesentery

Digestive tube

Coelomic cavity

Ventral mesentery

4 WEEKS

The digestive tube remains suspended in the coelom by a **dorsal mesentery** and a **ventral mesentery**. The ventral mesentery disintegrates everywhere except where major vessels or visceral organs have grown into it. It remains intact along the path of the *umbilical arteries* and where the *umbilical vein* and liver develop.

Pancreas

Greater omentum

Lesser omentum

Liver

As the embryo enlarges, the stomach and liver rotate toward the right, creating two pockets. The mesenteries that form these pockets are the greater omentum and the lesser omentum.

Pancreas

Coelom

Stomach

Falciform ligament

Liver

Neural tube

Aorta

Pancreas

Liver

The pancreas and liver begin as epithelial pockets that grow away from the digestive tract and into the dorsal and ventral mesenteries, respectively.

EMBRYOLOGY SUMMARY 19: THE DEVELOPMENT OF THE DIGESTIVE SYSTEM—*PART I*

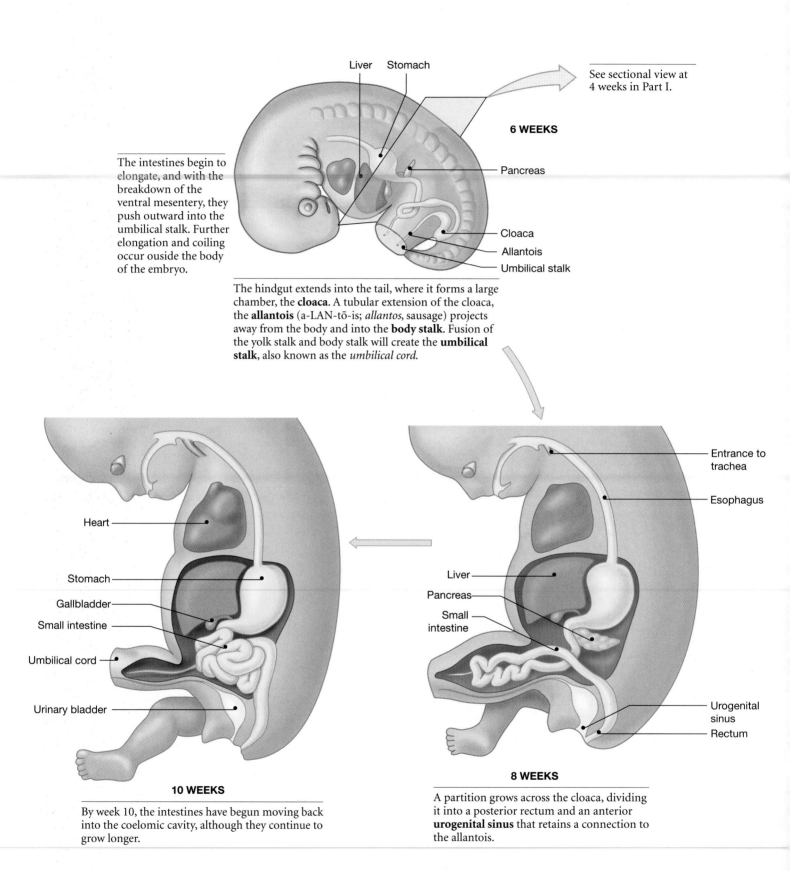

See sectional view at 4 weeks in Part I.

6 WEEKS

Liver   Stomach

Pancreas

Cloaca

Allantois

Umbilical stalk

The intestines begin to elongate, and with the breakdown of the ventral mesentery, they push outward into the umbilical stalk. Further elongation and coiling occur ouside the body of the embryo.

The hindgut extends into the tail, where it forms a large chamber, the **cloaca**. A tubular extension of the cloaca, the **allantois** (a-LAN-tō-is; *allantos*, sausage) projects away from the body and into the **body stalk**. Fusion of the yolk stalk and body stalk will create the **umbilical stalk**, also known as the *umbilical cord.*

Entrance to trachea

Esophagus

Liver

Pancreas

Small intestine

Urogenital sinus

Rectum

8 WEEKS

A partition grows across the cloaca, dividing it into a posterior rectum and an anterior **urogenital sinus** that retains a connection to the allantois.

Heart

Stomach

Gallbladder

Small intestine

Umbilical cord

Urinary bladder

10 WEEKS

By week 10, the intestines have begun moving back into the coelomic cavity, although they continue to grow longer.

EMBRYOLOGY SUMMARY 19: THE DEVELOPMENT OF THE DIGESTIVE SYSTEM—*PART II*

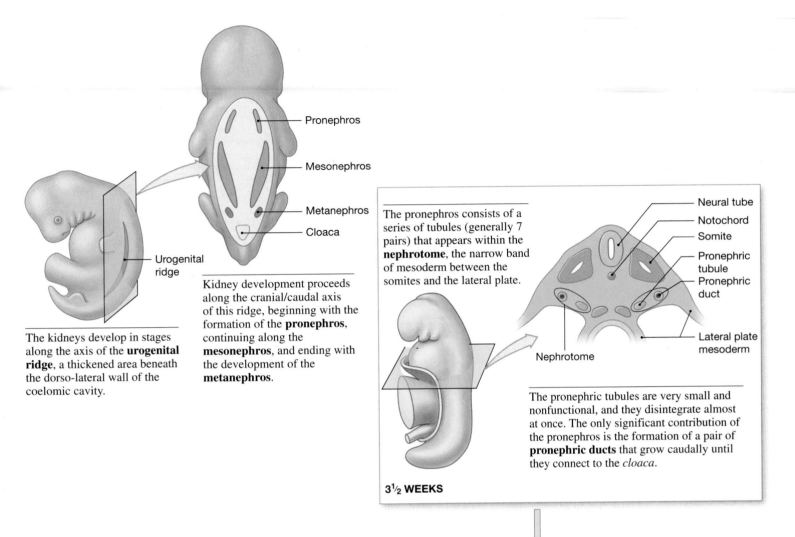

The kidneys develop in stages along the axis of the **urogenital ridge**, a thickened area beneath the dorso-lateral wall of the coelomic cavity.

Kidney development proceeds along the cranial/caudal axis of this ridge, beginning with the formation of the **pronephros**, continuing along the **mesonephros**, and ending with the development of the **metanephros**.

Pronephros

Mesonephros

Metanephros

Cloaca

Urogenital ridge

The pronephros consists of a series of tubules (generally 7 pairs) that appears within the **nephrotome**, the narrow band of mesoderm between the somites and the lateral plate.

Neural tube

Notochord

Somite

Pronephric tubule

Pronephric duct

Lateral plate mesoderm

Nephrotome

The pronephric tubules are very small and nonfunctional, and they disintegrate almost at once. The only significant contribution of the pronephros is the formation of a pair of **pronephric ducts** that grow caudally until they connect to the *cloaca*.

3½ WEEKS

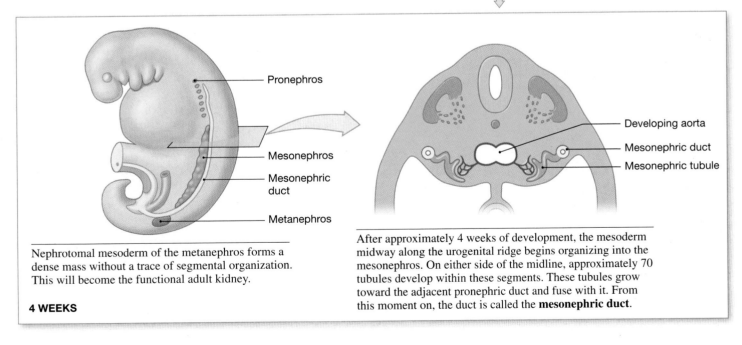

Pronephros

Mesonephros

Mesonephric duct

Metanephros

Nephrotomal mesoderm of the metanephros forms a dense mass without a trace of segmental organization. This will become the functional adult kidney.

4 WEEKS

Developing aorta

Mesonephric duct

Mesonephric tubule

After approximately 4 weeks of development, the mesoderm midway along the urogenital ridge begins organizing into the mesonephros. On either side of the midline, approximately 70 tubules develop within these segments. These tubules grow toward the adjacent pronephric duct and fuse with it. From this moment on, the duct is called the **mesonephric duct**.

EMBRYOLOGY SUMMARY 20: **THE DEVELOPMENT OF THE URINARY SYSTEM—PART I**

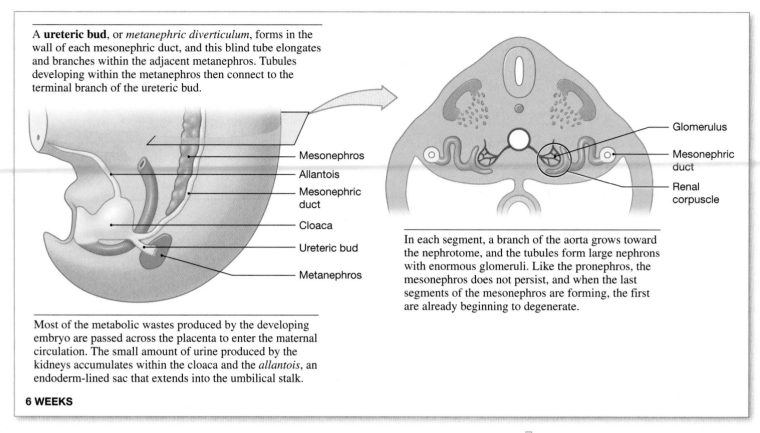

A **ureteric bud**, or *metanephric diverticulum*, forms in the wall of each mesonephric duct, and this blind tube elongates and branches within the adjacent metanephros. Tubules developing within the metanephros then connect to the terminal branch of the ureteric bud.

Mesonephros
Allantois
Mesonephric duct
Cloaca
Ureteric bud
Metanephros

Glomerulus
Mesonephric duct
Renal corpuscle

In each segment, a branch of the aorta grows toward the nephrotome, and the tubules form large nephrons with enormous glomeruli. Like the pronephros, the mesonephros does not persist, and when the last segments of the mesonephros are forming, the first are already beginning to degenerate.

Most of the metabolic wastes produced by the developing embryo are passed across the placenta to enter the maternal circulation. The small amount of urine produced by the kidneys accumulates within the cloaca and the *allantois*, an endoderm-lined sac that extends into the umbilical stalk.

6 WEEKS

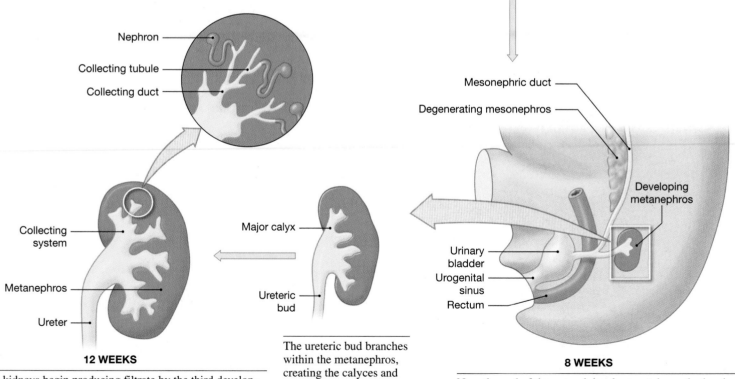

Nephron
Collecting tubule
Collecting duct

Mesonephric duct
Degenerating mesonephros
Developing metanephros

Collecting system
Metanephros
Ureter

Major calyx
Ureteric bud

Urinary bladder
Urogenital sinus
Rectum

12 WEEKS

The ureteric bud branches within the metanephros, creating the calyces and the collecting system. The nephrons, which form within the mesoderm of the metanephros, tap into the collecting tubules.

8 WEEKS

The kidneys begin producing filtrate by the third developmental month. The filtrate does not contain waste products, as they are excreted at the placenta for removal and elimination by the maternal kidneys. The sterile filtrate mixes with the amniotic fluid and is swallowed by the fetus and reabsorbed across the lining of the digestive tract.

Near the end of the second developmental month, the cloaca is subdivided into a dorsal rectum and a ventral **urogenital sinus**. The proximal portions of the allantois persist as the **urinary bladder**, and the connection between the bladder and an opening on the body surface will form the **urethra**.

EMBRYOLOGY SUMMARY **20**: THE DEVELOPMENT OF THE URINARY SYSTEM—*PART II*

SEXUALLY INDIFFERENT STAGES
(WEEKS 3–6)

DEVELOPMENT OF THE GONADS

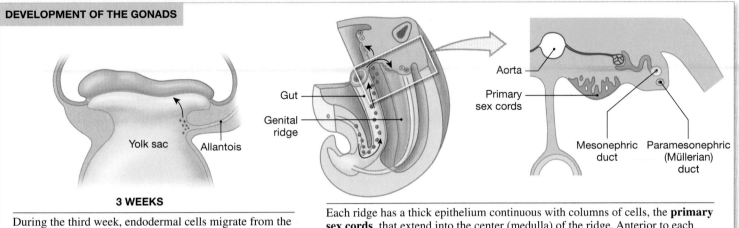

3 WEEKS

During the third week, endodermal cells migrate from the wall of the yolk sac near the allantois to the dorsal wall of the abdominal cavity. These primordial germ cells enter the **genital ridges** that parallel the mesonephros.

Each ridge has a thick epithelium continuous with columns of cells, the **primary sex cords**, that extend into the center (medulla) of the ridge. Anterior to each mesonephric duct, a duct forms that has no connection to the kidneys. This is the **paramesonephric** (*Müllerian*) **duct**; it extends along the genital ridge and continues toward the cloaca. At this sexually indifferent stage, male embryos cannot be distinguished from female embryos.

DEVELOPMENT OF DUCTS AND ACCESSORY ORGANS

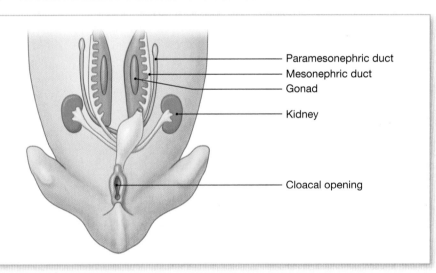

Both sexes have mesonephric and paramesonephric ducts at this stage. Unless exposed to androgens, the embryo—regardless of its genetic sex—will develop into a female. In a normal male embryo, cells in the core (medulla) of the genital ridge begin producing testosterone sometime after week 6. Testosterone triggers the changes in the duct system and external genitalia that are detailed on the following page.

DEVELOPMENT OF EXTERNAL GENITALIA

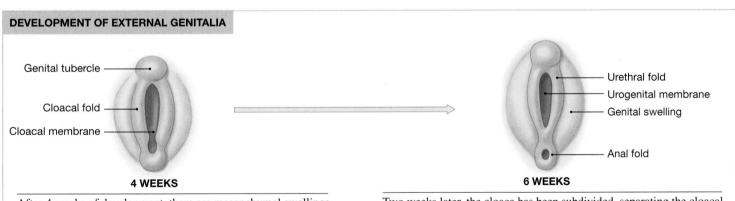

4 WEEKS

After 4 weeks of development, there are mesenchymal swellings called **cloacal folds** around the **cloacal membrane** (the cloaca does not open to the exterior). The **genital tubercle** forms the glans of the penis in males and the clitoris in females.

6 WEEKS

Two weeks later, the cloaca has been subdivided, separating the cloacal membrane into a posterior *anal membrane*, bounded by the *anal folds*, and an anterior **urogenital membrane**, bounded by the **urethral folds**. A prominent **genital swelling** forms lateral to each urethral fold.

DEVELOPMENT OF THE MALE REPRODUCTIVE SYSTEM

DEVELOPMENT OF THE TESTES

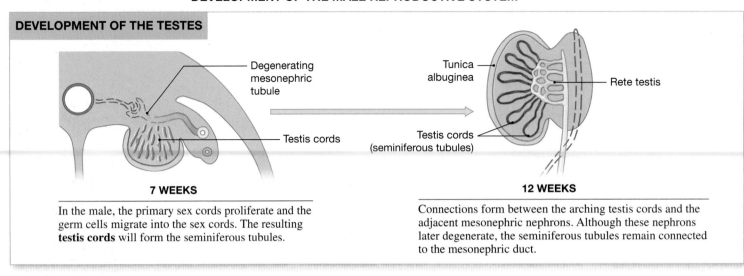

7 WEEKS

12 WEEKS

In the male, the primary sex cords proliferate and the germ cells migrate into the sex cords. The resulting **testis cords** will form the seminiferous tubules.

Connections form between the arching testis cords and the adjacent mesonephric nephrons. Although these nephrons later degenerate, the seminiferous tubules remain connected to the mesonephric duct.

DEVELOPMENT OF MALE DUCTS AND ACCESSORY ORGANS

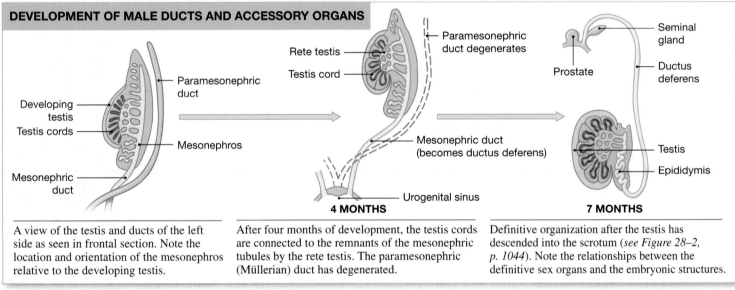

4 MONTHS

7 MONTHS

A view of the testis and ducts of the left side as seen in frontal section. Note the location and orientation of the mesonephros relative to the developing testis.

After four months of development, the testis cords are connected to the remnants of the mesonephric tubules by the rete testis. The paramesonephric (Müllerian) duct has degenerated.

Definitive organization after the testis has descended into the scrotum (*see Figure 28–2, p. 1044*). Note the relationships between the definitive sex organs and the embryonic structures.

DEVELOPMENT OF MALE EXTERNAL GENITALIA

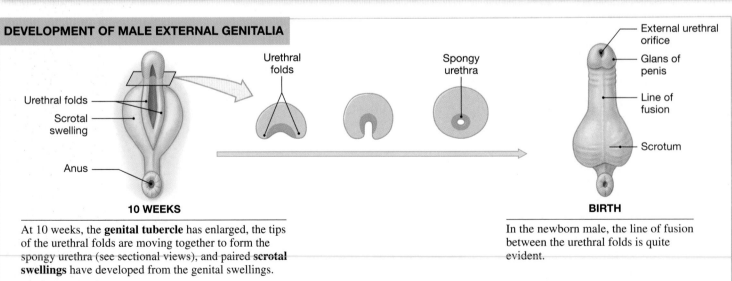

10 WEEKS

BIRTH

At 10 weeks, the **genital tubercle** has enlarged, the tips of the urethral folds are moving together to form the spongy urethra (see sectional views), and paired **scrotal swellings** have developed from the genital swellings.

In the newborn male, the line of fusion between the urethral folds is quite evident.

DEVELOPMENT OF THE FEMALE REPRODUCTIVE SYSTEM

DEVELOPMENT OF THE OVARIES

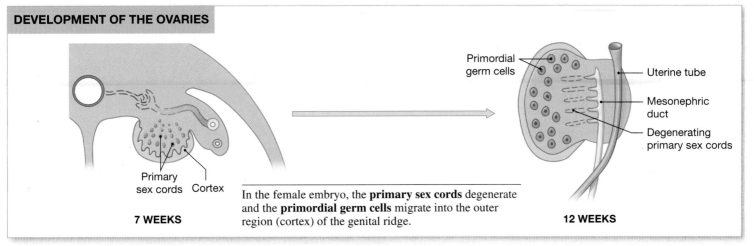

Primordial germ cells

Uterine tube

Mesonephric duct

Degenerating primary sex cords

Primary sex cords   Cortex

7 WEEKS

In the female embryo, the **primary sex cords** degenerate and the **primordial germ cells** migrate into the outer region (cortex) of the genital ridge.

12 WEEKS

DEVELOPMENT OF FEMALE DUCTS AND ACCESSORY ORGANS

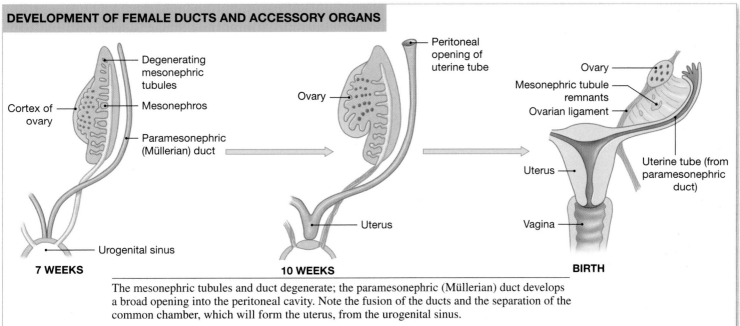

Degenerating mesonephric tubules

Mesonephros

Cortex of ovary

Paramesonephric (Müllerian) duct

Urogenital sinus

7 WEEKS

Peritoneal opening of uterine tube

Ovary

Uterus

10 WEEKS

Ovary

Mesonephric tubule remnants

Ovarian ligament

Uterus

Uterine tube (from paramesonephric duct)

Vagina

BIRTH

The mesonephric tubules and duct degenerate; the paramesonephric (Müllerian) duct develops a broad opening into the peritoneal cavity. Note the fusion of the ducts and the separation of the common chamber, which will form the uterus, from the urogenital sinus.

COMPARISON OF MALE AND FEMALE EXTERNAL GENITALIA

Males
Penis
  Corpora cavernosa
  Corpus spongiosum
  Proximal shaft of penis
  Spongy urethra
Bulbo-urethral glands
Scrotum

Females
Clitoris
  Erectile tissue
  Vestibular bulbs
  Labia minora
  Vestibule
Greater vestibular glands
Labia majora

DEVELOPMENT OF FEMALE EXTERNAL GENITALIA

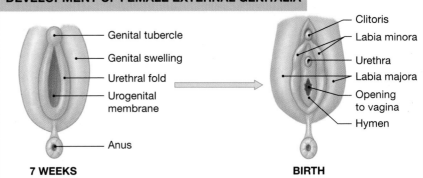

Genital tubercle

Genital swelling

Urethral fold

Urogenital membrane

Anus

7 WEEKS

Clitoris

Labia minora

Urethra

Labia majora

Opening to vagina

Hymen

BIRTH

In the female, the urethral folds do not fuse; they develop into the labia minora. The genital swellings will form the labia majora. The genital tubercle develops into the clitoris. The urethra opens to the exterior immediately posterior to the clitoris. The hymen remains as an elaboration of the urogenital membrane.